Basic Linear Algebra

Exercises and Solutions

Other World Scientific Titles by the Author

Introductory Topology: Exercises and Solutions
ISBN: 978-981-4583-81-7 (pbk)

Introductory Topology: Exercises and Solutions
Second Edition
ISBN: 978-981-314-693-8
ISBN: 978-981-314-802-4 (pbk)

An Operator Theory Problem Book
ISBN: 978-981-323-625-7

Basic Abstract Algebra: Exercises and Solutions
ISBN: 978-981-12-5210-5
ISBN: 978-981-12-5249-5 (pbk)

Basic Linear Algebra
Exercises and Solutions

Mohammed Hichem Mortad
University of Oran 1, Algeria

World Scientific

NEW JERSEY · LONDON · SINGAPORE · BEIJING · SHANGHAI · HONG KONG · TAIPEI · CHENNAI · TOKYO

Published by

World Scientific Publishing Co. Pte. Ltd.

5 Toh Tuck Link, Singapore 596224

USA office: 27 Warren Street, Suite 401-402, Hackensack, NJ 07601

UK office: 57 Shelton Street, Covent Garden, London WC2H 9HE

Library of Congress Cataloging-in-Publication Data

Names: Mortad, Mohammed Hichem, 1978–   author

Title: Basic linear algebra : exercises and solutions / Mohammed Hichem Mortad,
    University of Oran 1, Algeria.

Description: Singapore ; Hackensack, NJ : World Scientific Publishing Co. Pte. Ltd., [2026] |
    Includes bibliographical references and index.

Identifiers: LCCN 2025029988 | ISBN 9789811252679 hardcover |
    ISBN 9789811253379 paperback | ISBN 9789811252686 ebook for institutions |
    ISBN 9789811252693 ebook for individuals

Subjects: LCSH: Algebras, Linear--Textbooks

Classification: LCC QA184.2 .M67 2026

LC record available at https://lccn.loc.gov/2025029988

British Library Cataloguing-in-Publication Data

A catalogue record for this book is available from the British Library.

For any available supplementary material, please visit
https://www.worldscientific.com/worldscibooks/10.1142/12734#t=suppl

Contents

Introduction

This book is chiefly designed for first-year university students and instructors studying or teaching a fundamental linear algebra course. It offers a thorough introduction to linear algebra, organized into chapters that cover essential topics, from basic concepts to more advanced applications, with a focus on solved exercises for effective learning. The structure of the book includes four main sections in each chapter: *Course Summary*, *True or False*, *Exercises with Solutions*, and *Exercises without Solutions*. The book's design ensures that students can build their understanding progressively. Detailed solutions help reinforce the material, while the *True or False* section promotes critical thinking. Notice that while the majority of the questions in this section are framed as true or false, there are instances where certain questions do not have a simple binary answer. Some questions may require further exploration, where different cases or scenarios need to be discussed, depending on the context. Exercises without solutions encourage independent problem-solving, giving students the tools they need to assess their understanding and readiness for exams. This combination of theory, application, and self-assessment makes the book an ideal resource for both classroom learning and self-study.

While this book provides a concise yet solid foundation, it is not meant to serve as a substitute for a full course in linear algebra. To gain a deeper understanding of the subject, readers are encouraged to consult the many excellent books available. Some highly recommended texts include [8], [12], [13], [14], and [20].

The wide range of included exercises is drawn from various sources to provide students with ample practice opportunities. A fair amount of these exercises consists of standard and classical problems, sourced from materials such as [1], [2], [3], [4], [7], [9], [10], and [15]. I sincerely apologize for any inadvertent omissions in citations.

This book is the second in a series of three volumes designed for first-year students, each focusing on a different aspect of foundational mathematics. The other two volumes in the series are [18] and [19], and all three are being published by World Scientific.

I am deeply grateful to the team at World Scientific for their support. I would like to express my sincere gratitude to the staff for their patience and invaluable assistance throughout the production process. I am especially grateful to Miss Rosie Williams on her dedicated support.

As I conclude, I would like to invite readers to share their feedback, suggestions, or corrections to help improve future editions of this book. Please feel free to contact me via email at **mhmortad@gmail.com**.

Oran, 2025
Mohammed Hichem Mortad

CHAPTER 1

Matrices

1.1. Course Summary

1.1.1. Basic terminology.

DEFINITION 1.1.1. Let $\mathbb{F} = (\mathbb{F}, +, \cdot)$ be a field,[1] which is typically, but not exclusively, $\mathbb{R}$ or $\mathbb{C}$. Let $n, m \in \mathbb{N}$. A matrix (plural matrices) of size or dimension $n \times m$, with entries (or coefficients) in $\mathbb{F}$, is a table of n rows and m columns defined by

$$
A = \begin{pmatrix}
a_{11} & a_{12} & \cdots & a_{1m} \\
a_{21} & a_{22} & \cdots & a_{2m} \\
\cdots & \cdots & \cdots & \cdots \\
a_{n1} & a_{n2} & \cdots & a_{nm}
\end{pmatrix},
$$

where for all $1 \leq i \leq n$ and $1 \leq j \leq m$, $a_{ij} \in \mathbb{F}$. We may also write $A = (a_{ij})$.

The set of all such matrices is denoted by $M_{n \times m}(\mathbb{F})$, or merely by $M_{n \times m}$ when it is clear or unimportant to specify the underlying field $\mathbb{F}$.

Here are some special cases:

(1) When $n = m$, i.e., when the number of rows equals the number of columns, the matrix A is called a square matrix of size $n \times n$. A square matrix of size $n \times n$ is said to be a square matrix of order n. The set of square matrices of order n is denoted by $M_n(\mathbb{F})$ or simply M_n.

   Remark. Observe that $M_1(\mathbb{F})$, the space of 1×1 matrices over $\mathbb{F}$, is naturally identified with $\mathbb{F}$ itself.

(2) When $n = m = 1$, the matrix is reduced to an ordinary number.

(3) A row matrix is a matrix having one single row. Its size is of the type $1 \times m$. For example, $A = \begin{pmatrix} 1 & 0 & -1 \end{pmatrix}$ is a row matrix of size 1×3.

---

[1]In most algebra texts, a field is assumed to be commutative. We also adopt this assumption here.

(4) Similarly, a column matrix is a matrix having one single column. Its size is of the type $n \times 1$. For instance, $A = \begin{pmatrix} 1 \\ 2 \end{pmatrix}$ is a column matrix of size 2×1.

(5) A square matrix A of order n which is of the form

$$A = \begin{pmatrix} a_{11} & 0 & \cdots & 0 \\ a_{21} & a_{22} & \ddots & \vdots \\ \vdots & \ddots & \ddots & 0 \\ a_{n1} & \cdots & \cdots & a_{nn} \end{pmatrix}$$

is called a lower triangular matrix.

(6) An upper triangular matrix is a square matrix of the form

$$A = \begin{pmatrix} a_{11} & a_{12} & \cdots & a_{1n} \\ 0 & a_{22} & \ddots & a_{2n} \\ \vdots & \ddots & \ddots & \vdots \\ 0 & \cdots & 0 & a_{nn} \end{pmatrix}.$$

(7) A diagonal matrix is a square matrix of the form

$$A = \begin{pmatrix} a_{11} & 0 & \cdots & 0 \\ 0 & a_{22} & \ddots & \vdots \\ \vdots & \ddots & \ddots & 0 \\ 0 & \cdots & 0 & a_{nn} \end{pmatrix}.$$

(8) A particular diagonal matrix is when $a_{ii} = 1$ for each $i = 1, \cdots, n$. It is called the identity matrix, denoted by I_n, or merely by I when the size is clear or unimportant to specify. In other words,

$$I_n = \begin{pmatrix} 1 & 0 & \cdots & 0 \\ 0 & 1 & \ddots & \vdots \\ \vdots & \ddots & \ddots & 0 \\ 0 & \cdots & 0 & 1 \end{pmatrix}.$$

For example, $I_2 = \begin{pmatrix} 1 & 0 \\ 0 & 1 \end{pmatrix}$ and $I_3 = \begin{pmatrix} 1 & 0 & 0 \\ 0 & 1 & 0 \\ 0 & 0 & 1 \end{pmatrix}.$

(9) The zero (or null) matrix of size $n \times m$ is the matrix having zero entries only. It is designated by $0_{M_{n \times m}}$, or simply by 0 when its size is not essential to emphasize.

1.1.2. Matrix arithmetic: addition, multiplication, inversion, transposition and trace.

DEFINITION 1.1.2. Let $A = (a_{ij})$ and $B = (b_{ij})$ be two matrices.

(1) The matrices A and B are equal if they have the same size and their corresponding entries are equal, i.e., $a_{ij} = b_{ij}$ for all i, j.

(2) If A and B have the same size, their sum $A + B$ is obtained by adding the corresponding entries of A and B, i.e., $(A + B)_{ij} = a_{ij} + b_{ij}$.

(3) If α is a scalar, then the matrix αA is obtained by multiplying each entry of A by α, i.e., $(\alpha A)_{ij} = \alpha \cdot a_{ij}$.

(4) If the size of A is $n \times p$, and B is of size $p \times m$, then the product AB is a matrix of size $n \times m$, whose entry in row i and column j is obtained by calculating the inner (dot) product of the ith row of A with the jth column of B:

$$(AB)_{ij} = \sum_{k=1}^{p} a_{ik} b_{kj}.$$

(5) Let A be a square matrix and let $p \in \mathbb{N}$. We define the p power of A, denoted by A^p, by multiplying A with itself p times. In other words, A^p is defined as follows:

$$A^p = \underbrace{AA \cdots A}_{p \text{ times}}.$$

Additionally, we define:

$$A^0 = I, \ A^{p+1} = A^p A,$$

where I is the identity matrix.

Below are some of the arithmetic properties of matrix powers. Let $A \in M_n$, and let $p, q \in \mathbb{N}$. Then:

$$(A^p)^q = A^{pq} \quad \text{and} \quad A^{p+q} = A^p A^q.$$

(6) More generally, we may define $p(A)$, where A is a matrix and p is a polynomial. Let $A \in M_n(\mathbb{F})$. If $p(x) = a_0 + a_1 x + \cdots + a_p x^p$ is a polynomial, with $a_0, a_1, \ldots, a_p \in \mathbb{F}$, then we can define symbolically $p(A)$ as

$$p(A) = a_0 I + a_1 A + \cdots + a_n A^n,$$

where I is the identity matrix.

(7) A square matrix $A \in M_n$ is called nilpotent when $A^p = 0_{M_n}$ for some $p \in \mathbb{N}$.

Remark. Notice that the zero matrix is also considered nilpotent.

Next, we introduce two additional fundamental concepts.

DEFINITION 1.1.3. The transpose of a matrix $A = (a_{ij})_{1 \leq i \leq n, 1 \leq j \leq m}$ is the matrix denoted by A^t, obtained from A by converting its rows into columns and columns into rows. In other words, $A^t = (a_{ji})_{1 \leq j \leq m, 1 \leq i \leq n}$ (so, if $A \in M_{n \times m}$, then $A^t \in M_{m \times n}$).

DEFINITION 1.1.4. A matrix $A \in M_n(\mathbb{F})$ is said to be invertible if there exists a matrix $B \in M_n(\mathbb{F})$ such that $AB = BA = I_n$.

When such a B exists, which is known to be unique from general theory, it is called the inverse matrix of A, noted A^{-1}. In this case, $AA^{-1} = A^{-1}A = I_n$.

If matrix A is not invertible, we call it singular.

The following result is particularly useful when working with *square* matrices. It states that if A is a ***square*** matrix, then $AB = I$ (or $BA = I$) is sufficient for A to be invertible. A proof can be found in Exercise 3.3.34, as well as in the first remark below Theorem 4.1.1.

THEOREM 1.1.1. *Let $A \in M_n$. The following assertions are equivalent:*

(1) *A is invertible;*
(2) *There exists a $B \in M_n$ such that $BA = I$ (we then say that A is left invertible);*
(3) *There exists a $C \in M_n$ such that $AC = I$ (we then say that A is right invertible).*

Remark. In general rings with identity, the fact that $BA = I$ does not necessarily yield the invertibility of A. See, e.g., Exercise 3.3.33.

There are different methods for finding the inverse of an invertible matrix. Below, we present the method of Gaussian elimination.

- Let $A \in M_n$. To find A^{-1}, if it exists, we augment the matrix A with the identity matrix I_n of the same size and perform row operations (described below) to transform A into I_n. The matrix that results on the right side of the augmented matrix will be A^{-1}.

  Let us describe this method in more detail for the case $n = 3$. Form the augmented matrix $[A|I]$ as:

$$[A|I] = \begin{pmatrix} a_{11} & a_{12} & a_{13} & 1 & 0 & 0 \\ a_{21} & a_{22} & a_{23} & 0 & 1 & 0 \\ a_{31} & a_{32} & a_{33} & 0 & 0 & 1 \end{pmatrix}.$$

Use row operations (to both sides of the augmented matrix) to reduce A to the identity matrix. For example, if the matrix A is transformed into the identity matrix:

$$(A|I) \xrightarrow{\text{Row Operations}} \left(\begin{array}{ccc|ccc} 1 & 0 & 0 & b_{11} & b_{12} & b_{13} \\ 0 & 1 & 0 & b_{21} & b_{22} & b_{23} \\ 0 & 0 & 1 & b_{31} & b_{32} & b_{33} \end{array} \right),$$

then the right side of the augmented matrix will be A^{-1}. In other words,

$$A^{-1} = \left(\begin{array}{ccc} b_{11} & b_{12} & b_{13} \\ b_{21} & b_{22} & b_{23} \\ b_{31} & b_{32} & b_{33} \end{array} \right).$$

If, during the Gaussian elimination process, we are unable to find the identity matrix on the left (i.e., we encounter a row of all zeros), then this means that A is singular.

Finally, the allowed row operations used in Gaussian elimination are:

(1) Swap rows R_i and R_j:

$$R_i \leftrightarrow R_j$$

(2) Multiply row R_i by a nonzero scalar c:

$$R_i \to cR_i, \quad c \neq 0$$

(3) Add or subtract a multiple of one row to another:

$$R_j \to R_j - cR_i,$$

where c is a scalar.

We conclude this subsection by introducing two more important notions that readers will frequently encounter, both here and elsewhere. Their full significance will be explored in more detail later.

DEFINITION 1.1.5. Two $n \times n$ matrices A and B with entries from the same field are said to be similar if there exists an invertible $n \times n$ matrix T such that:

$$B = T^{-1}AT.$$

DEFINITION 1.1.6. The trace of a square matrix $A = (a_{ij}) \in M_n(\mathbb{F})$, denoted by $\text{tr}(A)$, is the sum of its diagonal entries:

$$\text{tr}(A) = \sum_{i=1}^{n} a_{ii}.$$

Remark. It is clear that if $A \in M_n(\mathbb{F})$, then $\text{tr}A \in \mathbb{F}$.

EXAMPLE 1.1.1. For the identity matrix I_n of size $n \times n$, we have:
$$\text{tr}(I_n) = n.$$

1.1.3. Basic properties of matrices.

Listed below are the basic algebraic properties of matrix addition, multiplication, trace, and transposition, assuming that all involved sums and products are well-defined (with brief details provided on a couple of occasions).

(1) Addition of Matrices:
- Commutativity: $A + B = B + A$ for matrices A and B of the same size.
- Associativity: $(A + B) + C = A + (B + C)$ for matrices A, B, C.
- Existence of zero matrix: There exists a zero matrix 0 such that $A + 0 = A$ for any matrix A.

(2) Scalar Multiplication:
- Distributivity (over addition of matrices): $\alpha(A + B) = \alpha A + \alpha B$, where α is a scalar.
- Distributivity (over scalar addition): $(\alpha + \beta)A = \alpha A + \beta A$, where α, β are scalars.
- Associativity: $\alpha(\beta A) = (\alpha\beta)A$, where α, β are scalars.
- Existence of identity: $1 \cdot A = A$, where 1 is the multiplicative identity scalar.

(3) Matrix Multiplication:
- Associativity: $A(BC) = (AB)C$ for matrices A, B, C (if the products are defined, that is, $A \in M_{n \times m}$, $B \in M_{m \times p}$ and $C \in M_{p \times q}$).
- Distributivity (over addition): The distributive properties of matrix multiplication are as follows:
  (a) $A(B+C) = AB+AC$, where $A \in M_{n \times m}$ and $B, C \in M_{m \times p}$.
  (b) $(A + B)C = AC + BC$, where $A, B \in M_{n \times m}$ and $C \in M_{m \times p}$.
- Non-commutativity: In general, matrix multiplication is *not commutative*, i.e., $AB \neq BA$ for matrices A and B (recall that two matrices A and B are said to commute provided $AB = BA$).
- Existence of identity matrix: There exists an identity matrix I (defined above) such that $AI = IA = A$ for any matrix A of compatible size.
- Multiplying by zero matrix: $A \cdot 0 = 0 \cdot A = 0$, where 0 is a zero matrix.

(4) Transpose of a Matrix:
- Transpose of a sum: $(A+B)^t = A^t + B^t$.
- Transpose of a scalar multiple: $(\alpha A)^t = \alpha A^t$.
- Double transpose: $(A^t)^t = A$.
- Transpose of a product: $(AB)^t = B^t A^t$.

(5) Inverse of a Matrix: Let $A, B \in M_n$ be two invertible matrices.
- A^{-1} is invertible and $(A^{-1})^{-1} = A$.
- $(AB)^{-1} = B^{-1}A^{-1}$.

(6) Trace of a Matrix:
- "Linearity" (See Exercise 3.3.9):

$$\operatorname{tr}(A+B) = \operatorname{tr}(A) + \operatorname{tr}(B) \quad \text{for all} \quad A, B \in M_n(\mathbb{F}).$$

$$\operatorname{tr}(\alpha A) = \alpha \operatorname{tr}(A) \quad \text{for any scalar } \alpha \in \mathbb{F}.$$

- Invariance under cyclic permutation: If A and B are matrices, square or not, as long as their product AB and BA are both defined, then

$$\operatorname{tr}(AB) = \operatorname{tr}(BA).$$

More generally, if A, B, and C are square matrices of appropriate sizes, then:

$$\operatorname{tr}(ABC) = \operatorname{tr}(CAB) = \operatorname{tr}(BCA).$$

1.2. True or False

Questions. Answer the following questions, providing justifications for your answers (A and B are two matrices in what follows):

(1) Give three reasons for which we, as a priori, do not have in general $AB = BA$, where A and B are two matrices.

(2) If $A + B$ is well-defined, then AB is well-defined.

(3) If $A + B$ is well-defined, then AB^t is well-defined.

(4) If AB and BA are well-defined, then $A + B$ is well-defined.

(5) Let A and B be two matrices such that their product AB is well-defined. Then

$$AB = 0 \Longrightarrow A = 0 \text{ or } B = 0.$$

(6) Let A, B be two matrices such that $AB = 0$. Then $BA = 0$.

(7) Let $A = \begin{pmatrix} 0 & 1 \\ -2 & 3 \end{pmatrix}$. Then $A^9 = \begin{pmatrix} 0 & 1 \\ (-2)^9 & 3^9 \end{pmatrix}$.

(8) Let $A \in M_n$ be such that $A^p = 0$ for a certain $p \in \mathbb{N}$ (with $n \geq 2$). Then $A = 0$.

(9) Let $A \in M_n$ be a diagonal matrix such that $A^p = 0$ for a certain $p \in \mathbb{N}$. Then $A = 0$.

(10) Let $A, B \in M_n$ and $p \in \mathbb{N}$, with $p \geq 2$. Then $(AB)^p = A^p B^p$.

(11) The product of upper triangular matrices (respectively lower) is always an upper triangular matrix (respectively lower).

(12) If A is a square matrix, then
$$A^2 \text{ diagonal} \implies A \text{ diagonal.}$$

(13) If A is invertible, then A^t is invertible.

(14) The zero matrix is invertible.

(15) The sum of two invertible matrices is always invertible.

(16) Each diagonal matrix is invertible.

(17) Let A be square matrix such that $A^3 = I$, where I is the identity matrix. Then A is invertible.

(18) Let A be a square matrix such that $A^3 = 0$, where 0 is the zero matrix. Then A is invertible.

(19) Let A and B be two invertible square matrices of the same order. Then AB is invertible and we have:
$$(AB)^{-1} = A^{-1}B^{-1}.$$

(20) If A is a nilpotent matrix, then so is A^t.

(21) Let A be an arbitrary matrix. Then $\text{tr}(A^t A)$ is always well-defined.

(22) Let $A, B \in M_n$. Then
$$\text{tr}(AB) = \text{tr}A \cdot \text{tr}B.$$

(23) Let $A, B, C \in M_n$. Then
$$\text{tr}(ABC) = \text{tr}(BAC).$$

(24) If A and B are similar matrices, then
$$\text{tr}A = \text{tr}B.$$

(25) Let $A, B \in M_n$. Fill in the gaps:
  (a) $(A + B)^2 = A^2 + B^2 + \cdots$.
  (b) $(A + B)(A - B) = \cdots$.
  (c) $\text{tr}[(A + B)^2] = \text{tr}(A^2) + \text{tr}(B^2) + \cdots$.
  (d) $\text{tr}[(A + B)(A - B)] = \text{tr}(A^2) - \text{tr}(B^2) + \cdots$.

Answers.

(1) The product of matrices is not commutative, meaning that the equality of AB and BA is not always guaranteed for two given matrices, A and B. There are several reasons for this:
  (a) First reason: The product AB may not be well-defined, while the product BA is well-defined. For example, if matrix A is of size 2×3 and matrix B is of size 3×4.

(b) Second reason: Both products, AB and BA, may be well-defined, but the resulting matrices may have different sizes. For instance, if A is of size 3×2 and B is of size 2×3, then AB is of size 3×3 while BA has size 2×2.

(c) Third reason: There may be matrices whose products are well-defined and of the same size, yet the products are not equal. For example, let:

$$A = \begin{pmatrix} 0 & 1 \\ 1 & 0 \end{pmatrix} \text{ and } B = \begin{pmatrix} 0 & 0 \\ 1 & 0 \end{pmatrix}.$$

Then

$$AB = \begin{pmatrix} 1 & 0 \\ 0 & 0 \end{pmatrix} \neq BA = \begin{pmatrix} 0 & 0 \\ 0 & 1 \end{pmatrix}.$$

(2) **False!** For instance, if A and B are of size 3×2, then $A + B$ is well-defined, but the product AB is not, as the number of columns in A does not match the number of rows in B.

(3) **True.** If $A + B$ is well-defined, then A and B must be of the same size, for example, of size $n \times p$. Therefore, B^t will be of size $p \times n$, and AB^t will be well-defined, with a resulting size of $n \times n$.

(4) **False!** For example, if A and B are of size 3×2 and 2×3, respectively, then AB is well-defined (its size is 3×3), and BA is well-defined (its size is 2×2). However, $A + B$ is not well-defined!

(5) **False!** For example, let

$$A = \begin{pmatrix} 1 & 0 \\ 0 & 0 \end{pmatrix} \text{ and } B = \begin{pmatrix} 0 & 0 \\ 0 & 1 \end{pmatrix}.$$

Then $A \neq 0_{M_2(\mathbb{R})}$ and $B \neq 0_{M_2(\mathbb{R})}$, yet

$$AB = \begin{pmatrix} 1 & 0 \\ 0 & 0 \end{pmatrix} \begin{pmatrix} 0 & 0 \\ 0 & 1 \end{pmatrix} = \begin{pmatrix} 0 & 0 \\ 0 & 0 \end{pmatrix} = 0_{M_2(\mathbb{R})}.$$

(6) **False!** For example, if $A = \begin{pmatrix} 0 & 1 \\ 0 & 0 \end{pmatrix}$ and $B = \begin{pmatrix} 1 & 0 \\ 0 & 0 \end{pmatrix}$, then $AB = \begin{pmatrix} 0 & 0 \\ 0 & 0 \end{pmatrix}$, while $BA = \begin{pmatrix} 0 & 1 \\ 0 & 0 \end{pmatrix} \neq \begin{pmatrix} 0 & 0 \\ 0 & 0 \end{pmatrix}.$

(7) **False!** To find A^9, we need to multiply A by itself nine times. Such calculations can be challenging, mainly when dealing with larger powers or bigger matrices. Thanks to Exercise 1.3.13, we have a general formula for A^n, where it suffices to set $n = 9$. However, this formula seems quite hard to derive,

even impossible in certain other cases. This is when we need to "diagonalize" A, and readers should wait until reaching Exercise 7.3.4 to see how we can obtain such a formula.

(8) **False!** (Cf. Exercises 5.3.4 & 5.3.12) For example, consider $A = \begin{pmatrix} 0 & 1 \\ 0 & 0 \end{pmatrix}$, which is nonzero, yet $A^p = \begin{pmatrix} 0 & 0 \\ 0 & 0 \end{pmatrix}$ for all $p \geq 2$.

(9) **True.** Let us provide a proof. Consider the following diagonal matrix:

$$A = \begin{pmatrix} a_{11} & 0 & \cdots & 0 \\ 0 & a_{22} & \ddots & \vdots \\ \vdots & \ddots & \ddots & 0 \\ 0 & \cdots & 0 & a_{nn} \end{pmatrix}.$$

It can then be checked using a proof by induction that for any positive integer p:

$$A^p = \begin{pmatrix} a_{11}^p & 0 & \cdots & 0 \\ 0 & a_{22}^p & \ddots & \vdots \\ \vdots & \ddots & \ddots & 0 \\ 0 & \cdots & 0 & a_{nn}^p \end{pmatrix}.$$

Since $A^p = 0$, all entries of A^p must vanish, leading, in particular, to having $a_{ii}^p = 0$ for all $i = 1, \ldots, n$. Thus, $a_{ii} = 0$ for all $i = 1, \ldots, n$, or equivalently $A = 0$, as wished.

(10) **False!** For a counterexample, take $A = \begin{pmatrix} 0 & 1 \\ 0 & 0 \end{pmatrix}$ and $B = \begin{pmatrix} 0 & 0 \\ 1 & 0 \end{pmatrix}$. Then $A^p = B^p = 0$ for all $p \geq 2$. However,

$$AB = \begin{pmatrix} 0 & 1 \\ 0 & 0 \end{pmatrix} \begin{pmatrix} 0 & 0 \\ 1 & 0 \end{pmatrix} = \begin{pmatrix} 1 & 0 \\ 0 & 0 \end{pmatrix},$$

implying that $(AB)^p = \begin{pmatrix} 1 & 0 \\ 0 & 0 \end{pmatrix}$ for all $p \geq 2$. Thus, $(AB)^p \neq A^p B^p$ for all $p \geq 2$.

Remark. If $A, B \in M_n$ are such that $AB = BA$, then $(AB)^p = A^p B^p$ for all $p \geq 2$. Readers are referred to Exercise 1.3.8 for a proof.

(11) **True.** It is a very practical result, whose proof is left to the reader.

(12) **False!** Let $A = \begin{pmatrix} 1 & 2 \\ 3 & -1 \end{pmatrix}$. Then, A is not diagonal, and yet

$$A^2 = \begin{pmatrix} 7 & 0 \\ 0 & 7 \end{pmatrix}$$

is diagonal!

(13) **True.** If A is invertible, then:

$$\exists B : BA = AB = I.$$

Taking the transpose, we obtain:

$$(BA)^t = A^t B^t = (AB)^t = B^t A^t = I^t = I,$$

i.e. A^t is invertible, and its inverse is B^t.

(14) **False!** It is never invertible (as otherwise we would have $0 = I$, where 0 is the zero matrix); even a square matrix having only one row (or only one column) containing zeros only is never invertible!

(15) **False!** For example, if A is an invertible square matrix, then $-A$ is also invertible and of the same order, yet

$$A + (-A) = A - A = 0,$$

where 0 is the zero matrix, is not invertible.

The answer is even negative over $\mathbb{R}$, which is another way of viewing 1×1 real matrices. For example, let $A = (1)$ (a 1×1 matrix), and set $B = -A = (-1)$. Clearly, both A and B are invertible, yet their sum, $A + B = (0)$, is not invertible.

Remark. Students should not feel uncomfortable about these 1×1 matrices.

Remark. In the above counterexample, A commutes with B, a condition not readily available when dealing with matrices. However, this condition was not sufficient to make $A + B$ invertible.

(16) **False!** For instance,

$$A = \begin{pmatrix} 1 & 0 & 0 \\ 0 & 2 & 0 \\ 0 & 0 & 0 \end{pmatrix}$$

is a non-invertible diagonal matrix.

If A is a diagonal matrix and all the diagonal elements are *nonzero*, then A will be invertible. In this case, the inverse is also a diagonal matrix, where the entries on the diagonal are the reciprocals of the diagonal elements of the original matrix. In other words, if

$$A = \begin{pmatrix} a_1 & 0 & \cdots & 0 \\ 0 & a_2 & \ddots & \vdots \\ \vdots & \ddots & \ddots & 0 \\ 0 & \cdots & 0 & a_n \end{pmatrix},$$

where all a_i $(i = 1, \ldots, n)$ are nonzero, then A is invertible and we have:

$$A^{-1} = \begin{pmatrix} \frac{1}{a_1} & 0 & \cdots & 0 \\ 0 & \frac{1}{a_2} & \ddots & \vdots \\ \vdots & \ddots & \ddots & 0 \\ 0 & \cdots & 0 & \frac{1}{a_n} \end{pmatrix}.$$

Remark. If A is a triangular matrix (upper or lower) and the diagonal elements are all *nonzero*, then A will be invertible. In this case, the inverse of an upper triangular matrix (respectively lower triangular) is also an upper triangular matrix (respectively lower triangular).

(17) **True.** To show that A is invertible, and given its square form, we have to find a matrix B such that $AB = I$, where I is the usual identity matrix. Since

$$A^3 = I \iff A \underbrace{A^2}_{B} = I,$$

we see that A is invertible, with inverse given by $A^{-1} = A^2$.

(18) **False!** Let I be the identity matrix and 0 be the null matrix. Suppose, for contradiction, that A is invertible, and let B be its inverse. Then we have $AB = I$. From the hypothesis, we also have

$$A^2 AB = A^3 B = 0B = 0,$$

and since $AB = I$, it follows that

$$A^2 AB = A^2 I = A^2 = 0.$$

Similarly, from $A^2 = 0$, we obtain $A = 0$, which is a contradiction, since the null matrix is never invertible. Therefore, A cannot be invertible.

(19) **False!** While AB is invertible, it does not necessarily follow that $(AB)^{-1} = A^{-1}B^{-1}$, except in very particular cases, such

as when the matrices commute. However, recall that the correct formula:

$$(AB)^{-1} = B^{-1}A^{-1}$$

(if both A and B are invertible).

A related result is as follows, and its proof is relatively straightforward.

THEOREM. *If A and B are square matrices of the same order such that AB is invertible, then both A and B are also invertible.*

(20) **True.** Since A is nilpotent, $A^p = 0$ for some $p \in \mathbb{N}$. We can prove by induction that $(A^p)^t = (A^t)^p$, from which we conclude that A^t is also nilpotent.

(21) **True.** To take the trace of a given matrix, it must be a square matrix, and this is the case for the matrix $A^t A$. Indeed, if A is an $n \times m$ matrix, then A^t is an $m \times n$ matrix, so $A^t A$ is well-defined and is a square matrix of order m. Therefore, we can compute its trace.

(22) **False!** For instance, if

$$A = \begin{pmatrix} 1 & 0 \\ 0 & 0 \end{pmatrix} \text{ and } B = \begin{pmatrix} 0 & 0 \\ 0 & 1 \end{pmatrix},$$

then $AB = \begin{pmatrix} 0 & 0 \\ 0 & 0 \end{pmatrix}$. Therefore,

$$\operatorname{tr}(AB) = 0 \neq 1 = 1 \cdot 1 = \operatorname{tr}A \cdot \operatorname{tr}B.$$

(23) **False!** The cyclic property of the trace must be respected, as it does not hold for arbitrary permutations. Specifically, we have the identity:

$$\operatorname{tr}(ABC) = \operatorname{tr}(CAB) = \operatorname{tr}(BCA),$$

but generally $\operatorname{tr}(ABC) \neq \operatorname{tr}(BAC)$.

Let us corroborate this doubt with a counterexample. Take

$$A = \begin{pmatrix} 0 & 1 \\ 0 & 0 \end{pmatrix}, \quad B = \begin{pmatrix} 0 & 0 \\ 1 & 0 \end{pmatrix} \text{ and } C = \begin{pmatrix} 1 & 0 \\ 0 & 0 \end{pmatrix}.$$

Then,

$$ABC = \begin{pmatrix} 0 & 1 \\ 0 & 0 \end{pmatrix}\begin{pmatrix} 0 & 0 \\ 1 & 0 \end{pmatrix}\begin{pmatrix} 1 & 0 \\ 0 & 0 \end{pmatrix} = \begin{pmatrix} 1 & 0 \\ 0 & 0 \end{pmatrix}\begin{pmatrix} 1 & 0 \\ 0 & 0 \end{pmatrix}$$

$$= \begin{pmatrix} 1 & 0 \\ 0 & 0 \end{pmatrix},$$

whilst

$$BAC = \begin{pmatrix} 0 & 0 \\ 1 & 0 \end{pmatrix} \begin{pmatrix} 0 & 1 \\ 0 & 0 \end{pmatrix} \begin{pmatrix} 1 & 0 \\ 0 & 0 \end{pmatrix} = \begin{pmatrix} 0 & 0 \\ 0 & 1 \end{pmatrix} \begin{pmatrix} 1 & 0 \\ 0 & 0 \end{pmatrix}$$

$$= \begin{pmatrix} 0 & 0 \\ 0 & 0 \end{pmatrix}.$$

So,
$$\text{tr}(ABC) = 1 \neq 0 = \text{tr}(BAC).$$

(24) **True.** Since A and B are similar matrices, $B = T^{-1}AT$ for some invertible matrix T. Then

$$\text{tr} B = \text{tr}(T^{-1}AT) = \text{tr}(TT^{-1}A) = \text{tr}(IA) = \text{tr}A,$$

as wished.

(25) (a) $(A + B)^2 = A^2 + B^2 + \cdots$: We have

$$(A + B)(A + B) = A^2 + AB + BA + B^2.$$

If $AB = BA$ (which is not always true), then

$$(A + B)^2 = A^2 + 2AB + B^2.$$

(b) $(A + B)(A - B) = \cdots$: Similarly,

$$(A + B)(A - B) = A^2 - AB + BA - B^2.$$

When $AB = BA$, this simplifies to

$$(A + B)(A - B) = A^2 - B^2.$$

(c) $\text{tr}[(A+B)^2] = \text{tr}(A^2)+\text{tr}(B^2)+\cdots$: Recall that $\text{tr}(AB) = \text{tr}(BA)$ always holds. Thus,

$$\text{tr}[(A + B)^2] = \text{tr}(A^2) + \text{tr}(AB) + \text{tr}(BA) + \text{tr}(B^2)$$
$$= \text{tr}(A^2) + 2\text{tr}(AB) + \text{tr}(B^2).$$

Consequently,

$$\text{tr}[(A + B)^2] = \text{tr}(A^2 + 2AB + B^2).$$

Remark. Although $(A+B)^2 \neq A^2+2AB+B^2$ in general, applying the trace makes it behave as if commutativity holds. This observation is crucial when using the trace in proofs.

(d) $\text{tr}[(A + B)(A - B)] = \text{tr}(A^2) - \text{tr}(B^2) + \cdots$: Using the trace property,

$$\text{tr}[(A + B)(A - B)] = \text{tr}(A^2 - AB + BA - B^2)$$
$$= \text{tr}(A^2 - B^2).$$

(Here again, the trace exhibits a commutativity-like behavior.)

1.3. Exercises with Solutions

Exercise 1.3.1. Find, when possible, AB, BA, and $A + B$, where
$$A = \begin{pmatrix} -1 & 2 \\ 0 & 1 \\ -2 & 0 \end{pmatrix}, \quad B = \begin{pmatrix} 1 & 0 & -1 \\ 3 & -1 & 2 \\ 1 & 1 & 0 \end{pmatrix}.$$

Solution 1.3.1. The product AB is not well-defined because A is a matrix of size 3×2 and B is of size 3×3. To observe why, note that matrix multiplication requires the number of columns of the first matrix to match the number of rows of the second. In this case:
$$3 \times \underbrace{2 \; 3}_{2 \neq 3} \times 3,$$
so AB is not possible.

However, BA is well-defined because B is of size 3×3 and A is of size 3×2. This time, the number of columns of B does match the number of rows of A. Specifically:
$$3 \times \underbrace{3 \; 3}_{3 = 3} \times 2 = \text{a matrix of size } (3 \times 2).$$

Thus, the product matrix BA will be of size 3×2, and we can compute:
$$BA = \begin{pmatrix} 1 & 0 & -1 \\ 3 & -1 & 2 \\ 1 & 1 & 0 \end{pmatrix} \begin{pmatrix} -1 & 2 \\ 0 & 1 \\ -2 & 0 \end{pmatrix}$$
$$= \begin{pmatrix} -1+0+2 & 2+0-0 \\ -3-0-4 & 6-1+0 \\ -1+0-0 & 2+1+0 \end{pmatrix} = \begin{pmatrix} 1 & 2 \\ -7 & 5 \\ -1 & 3 \end{pmatrix}.$$

Finally, the sum $A + B$ is not well-defined because A and B are not of the same size.

Exercise 1.3.2. Consider the same question as in the previous exercise:
$$A = \begin{pmatrix} 1 & 0 \\ 2 & -1 \end{pmatrix}, \quad B = \begin{pmatrix} 0 & -1 \\ 1 & 2 \end{pmatrix}.$$

Solution 1.3.2. All the operations requested are well-defined, and we have:
$$AB = \begin{pmatrix} 0 & -1 \\ -1 & -4 \end{pmatrix}, \quad BA = \begin{pmatrix} -2 & 1 \\ 5 & -2 \end{pmatrix},$$
and
$$A + B = \begin{pmatrix} 1 & -1 \\ 3 & 1 \end{pmatrix}.$$

Exercise 1.3.3. Consider the following matrices:
$$A = \begin{pmatrix} 0 & -2 & -1 \\ -3 & 1 & -1 \\ 0 & 2 & 2 \end{pmatrix} \text{ and } B = \begin{pmatrix} 2 & -2 & 4 \\ 0 & 1 & 3 \\ 1 & 1 & 0 \end{pmatrix}.$$

Find:
  (1) the first row of AB,
  (2) the third column of BA,
  (3) the diagonal of AB,
  (4) the second row of B^2.

Solution 1.3.3. Readers can compute the products AB, BA, and B^2, and then answer the questions. While this is not incorrect, it is unnecessary.

  (1) To determine the first row of AB, it suffices to compute the dot product of the first row of A with each column of B, and we get:
$$\begin{pmatrix} 0-0-1 & -0-2-1 & 0-6-0 \end{pmatrix} = \begin{pmatrix} -1 & -3 & -6 \end{pmatrix}.$$

  (2) To determine the third column of BA, we compute the dot product of the rows of B with the third column of A. We obtain:
$$\begin{pmatrix} -2+2+8 \\ -0-1+6 \\ -1-1+0 \end{pmatrix} = \begin{pmatrix} 8 \\ 5 \\ -2 \end{pmatrix}.$$

  (3) To find the diagonal of AB, we compute the dot product of the elements of the first row of A with the corresponding elements of the first column of B, then repeat the process for the second row with the second column and the third row with the third column. This gives us:
$$\begin{pmatrix} -1 & & \\ & 6 & \\ & & 6 \end{pmatrix}.$$

  (4) Finally, to find the second row of B^2, we compute the dot product of the second row of B with each column of B. This gives us:
$$\begin{pmatrix} 3 & 4 & 3 \end{pmatrix}.$$

Exercise 1.3.4. Find the transpose of the following matrices:
$$A = \begin{pmatrix} -1 & 2 \\ 1 & 0 \end{pmatrix}, \quad B = \begin{pmatrix} 2 & \pi \\ 0 & -1 \\ -4 & 1 \end{pmatrix}, \quad C = \begin{pmatrix} 1 & -1 & 0 \\ -1 & -2 & 2 \\ 0 & 2 & 5 \end{pmatrix}.$$

Solution 1.3.4. To find the transpose of a matrix, recall that each row becomes a column and each column becomes a row. Thus, we have:

$$A^t = \begin{pmatrix} -1 & 1 \\ 2 & 0 \end{pmatrix}, \ B^t = \begin{pmatrix} 2 & 0 & -4 \\ \pi & -1 & 1 \end{pmatrix}, \ C^t = \begin{pmatrix} 1 & -1 & 0 \\ -1 & -2 & 2 \\ 0 & 2 & 5 \end{pmatrix}$$

(observe that we have obtained $C^t = C$).

Exercise 1.3.5. Find a *diagonal* matrix $A \in M_3(\mathbb{R})$ such that

$$A^5 = \begin{pmatrix} 1 & 0 & 0 \\ 0 & -1 & 0 \\ 0 & 0 & -1 \end{pmatrix}.$$

Solution 1.3.5. Since A is diagonal, it is written as:

$$\begin{pmatrix} a & 0 & 0 \\ 0 & b & 0 \\ 0 & 0 & c \end{pmatrix},$$

where $a, b, c \in \mathbb{R}$. Then, we have:

$$A^5 = \begin{pmatrix} a^5 & 0 & 0 \\ 0 & b^5 & 0 \\ 0 & 0 & c^5 \end{pmatrix}.$$

For A^5 to equal the given matrix, we must have:

$$a^5 = 1, \ b^5 = -1 \text{ and } c^5 = -1,$$

so $a = 1$ and $b = c = -1$ (since we are looking for real solutions only).

Exercise 1.3.6. Solve the following matrix equation:

$$\begin{pmatrix} a-b & b+c \\ 3d+c & 2a-4d \end{pmatrix} = \begin{pmatrix} 8 & 1 \\ 7 & 6 \end{pmatrix},$$

where $a, b, c, d \in \mathbb{R}$.

Solution 1.3.6. The unknowns a, b, c, and d satisfy the following system:

$$\begin{cases} a - b = 8, \\ b + c = 1, \\ 3d + c = 7, \\ 2a - 4d = 6. \end{cases}$$

We find

$$a = 5, \ b = -3, c = 4, \text{ and } d = 1.$$

Exercise 1.3.7. Let $a, b \in \mathbb{R}^*$. Let

$$A = \begin{pmatrix} a & b \\ 0 & a \end{pmatrix}.$$

Give all square matrices of order 2 commuting with A.

Solution 1.3.7. Let $X = \begin{pmatrix} x & y \\ z & t \end{pmatrix}$ be a matrix commuting with A, i.e., $AX = XA$, i.e.,

$$\begin{pmatrix} a & b \\ 0 & a \end{pmatrix} \begin{pmatrix} x & y \\ z & t \end{pmatrix} = \begin{pmatrix} x & y \\ z & t \end{pmatrix} \begin{pmatrix} a & b \\ 0 & a \end{pmatrix},$$

i.e.,

$$\begin{pmatrix} ax + bz & ay + bt \\ az & at \end{pmatrix} = \begin{pmatrix} xa & xb + ya \\ za & zb + ta \end{pmatrix}.$$

Then,

$$\begin{cases} ax + bz = xa, \\ ay + bt = xb + ya, \\ za = az, \\ zb + ta = at. \end{cases}$$

So,

$$\begin{cases} bz = 0, \\ bt = xb, \end{cases}$$

and since $a, b \neq 0$, we get $z = 0$, $x = t$, and y is arbitrary. Therefore, the matrices commuting with A are of the form:

$$\begin{pmatrix} x & y \\ 0 & x \end{pmatrix},$$

where x and y are real scalars.

Exercise 1.3.8. Let $A, B \in M_n$ be such that $AB = BA$.
(1) Show that $AB^p = B^p A$ for all $p \in \mathbb{N}$. Hence, $A^q B^p = B^p A^q$ for all $p \in \mathbb{N}$ and for all $q \in \mathbb{N}$.
(2) Show that $(AB)^p = A^p B^p = B^p A^p$ for all $p \in \mathbb{N}$.

Solution 1.3.8.

(1) We use a proof by induction. Since $AB = BA$, the statement is true for $p = 1$, which is enough for the base case, but it is more enlightening to check its trueness for $p = 2$. Since $AB = BA$, it follows, using the associativity of matrix multiplication, that

$$AB^2 = ABB = (AB)B = (BA)B = B(AB) = B(BA) = B^2 A.$$

Now, assume $AB^p = B^p A$. Then

$$AB^{p+1} = AB^p B = B^p AB = B^p BA = B^{p+1} A.$$

Thus, $AB^p = B^p A$ for all $p \in \mathbb{N}$. By a similar reasoning, we may show that $A^q B^p = B^p A^q$ for all $q \in \mathbb{N}$ (and $p \in \mathbb{N}$).

(2) Here again, we use a proof by induction, and the base case is trivial. So, suppose $(AB)^p = A^p B^p$. Hence

$$(AB)^{p+1} = (AB)^p AB = A^p B^p AB = A^p AB^p B = A^{p+1} B^{p+1}.$$

Therefore, we can conclude that $(AB)^p = A^p B^p$ for all $p \in \mathbb{N}$. Using the previous question, we can also conclude that $(AB)^p = (BA)^p = B^p A^p$ for all $p \in \mathbb{N}$.

Exercise 1.3.9. Let $A, B \in M_n(\mathbb{R})$ be such that $A^2 B^2 = B^2 A^2$. Does it follow that $AB = BA$?

Solution 1.3.9. The answer is negative. Indeed, consider $A = \begin{pmatrix} 0 & 1 \\ 0 & 0 \end{pmatrix}$, then $A^2 = 0_{M_2(\mathbb{R})}$, which means that $A^2 B^2 = B^2 A^2$, even without having to compute B^2. It remains to choose a $B \in M_2(\mathbb{R})$ that does not commute with A. For instance, if $B = \begin{pmatrix} 0 & 0 \\ 1 & 0 \end{pmatrix}$, then

$$AB = \begin{pmatrix} 1 & * \\ * & * \end{pmatrix} \neq \begin{pmatrix} 0 & * \\ * & * \end{pmatrix} = AB,$$

where we did not need to calculate all entries, as we only had to find two different values lying on the same position in each of AB and BA.

Exercise 1.3.10. Let $D \in M_n$ be diagonal and let $p(t)$ be a polynomial of degree m with $p(t) = a_0 + a_1 t + \cdots + a_m t^m$. Show that $p(D)$ is diagonal.

Solution 1.3.10. Since D is diagonal, it has the form

$$D = \begin{pmatrix} \alpha_{11} & 0 & \cdots & 0 \\ 0 & \alpha_{22} & \ddots & \vdots \\ \vdots & \ddots & \ddots & 0 \\ 0 & \cdots & 0 & \alpha_{nn} \end{pmatrix}.$$

Then, for all $m \in \mathbb{N}$, we have

$$D^m = \begin{pmatrix} \alpha_{11}^m & 0 & \cdots & 0 \\ 0 & \alpha_{22}^m & \ddots & \vdots \\ \vdots & \ddots & \ddots & 0 \\ 0 & \cdots & 0 & \alpha_{nn}^m \end{pmatrix}.$$

Given the polynomial expression $p(D) = a_0 I + a_1 D + \cdots + a_m D^m$, it can be observed that $p(D)$ can be represented as a diagonal matrix:

$$p(D) = \begin{pmatrix} p(\alpha_{11}) & 0 & \cdots & 0 \\ 0 & p(\alpha_{22}) & \ddots & \vdots \\ \vdots & \ddots & \ddots & 0 \\ 0 & \cdots & 0 & p(\alpha_{nn}) \end{pmatrix},$$

which signifies that $p(D)$ is indeed a diagonal matrix, as required.

Exercise 1.3.11. Is the sum of two nilpotent matrices nilpotent? What about their product?

Solution 1.3.11. Let

$$A = \begin{pmatrix} 0 & 1 \\ 0 & 0 \end{pmatrix} \text{ and } B = \begin{pmatrix} 0 & 0 \\ 1 & 0 \end{pmatrix}.$$

Then, both A and B are nilpotent while $A + B$ is not nilpotent as

$$(A + B)^2 = \begin{pmatrix} 1 & 0 \\ 0 & 1 \end{pmatrix} \neq \begin{pmatrix} 0 & 0 \\ 0 & 0 \end{pmatrix}$$

(in fact, $(A + B)^p \neq 0$ for all $p \geq 2$). The same pair of matrices serves as a counterexample for the product. Indeed,

$$AB = \begin{pmatrix} 0 & 1 \\ 0 & 0 \end{pmatrix} \begin{pmatrix} 0 & 0 \\ 1 & 0 \end{pmatrix} = \begin{pmatrix} 1 & 0 \\ 0 & 0 \end{pmatrix}$$

is obviously not nilpotent since $(AB)^p \neq 0$ for all $p \geq 2$.

Exercise 1.3.12. Let $A, B \in M_n$ be such that AB is nilpotent. Show that BA is nilpotent.

Solution 1.3.12. Since AB is nilpotent, we know that $(AB)^p = 0$ for a certain $p \in \mathbb{N}$. Hence

$$(BA)^{p+1} = \underbrace{BABA \cdots BA}_{p+1 \text{ times}} = B(\underbrace{ABA \cdots B}_{p \text{ times}})A = B(AB)^p A = B0A = 0,$$

indicating that BA is nilpotent.

Exercise 1.3.13. (Cf. Exercise 7.3.4) Let $A = \begin{pmatrix} 0 & 1 \\ -2 & 3 \end{pmatrix}$. Using a proof by induction, show that

$$A^n = \begin{pmatrix} 2 - 2^n & 2^n - 1 \\ 2 - 2^{n+1} & 2^{n+1} - 1 \end{pmatrix}$$

for all $n \in \mathbb{N}$.

Solution 1.3.13. The base case is true because when $n = 1$, we have

$$A^1 = \begin{pmatrix} 2 - 2^1 & 2^1 - 1 \\ 2 - 2^{1+1} & 2^{1+1} - 1 \end{pmatrix} = \begin{pmatrix} 2 - 2 & 2 - 1 \\ 2 - 4 & 4 - 1 \end{pmatrix} = \begin{pmatrix} 0 & 1 \\ -2 & 3 \end{pmatrix} = A.$$

Now, suppose $A^n = \begin{pmatrix} 2 - 2^n & 2^n - 1 \\ 2 - 2^{n+1} & 2^{n+1} - 1 \end{pmatrix}$. We then have

$$
\begin{aligned}
A^{n+1} = A^n A &= \begin{pmatrix} 2 - 2^n & 2^n - 1 \\ 2 - 2^{n+1} & 2^{n+1} - 1 \end{pmatrix} \begin{pmatrix} 0 & 1 \\ -2 & 3 \end{pmatrix} \\
&= \begin{pmatrix} -2^{n+1} + 2 & 2 - 2^n + 3 \times 2^n - 3 \\ -2 \times 2^{n+1} + 2 & 2 - 2^{n+1} + 3 \times 2^{n+1} - 3 \end{pmatrix} \\
&= \begin{pmatrix} 2 - 2^{n+1} & 2^{n+1} - 1 \\ 2 - 2^{n+2} & 2^{n+2} - 1 \end{pmatrix},
\end{aligned}
$$

which is the expected formula for A^{n+1}.

Exercise 1.3.14. Let

$$A = \begin{pmatrix} 2 & 2 & -1 \\ 0 & 2 & 2 \\ 0 & 0 & 2 \end{pmatrix}.$$

(1) Find a square matrix B such that $A = 2I + B$, where I is the identity matrix of order 3.
(2) Find B^2 and B^3.
(3) Deduce A^n for all $n \in \mathbb{N}$.

Solution 1.3.14.

(1) We can write:

$$\begin{pmatrix} 2 & 2 & -1 \\ 0 & 2 & 2 \\ 0 & 0 & 2 \end{pmatrix} = \underbrace{\begin{pmatrix} 2 & 0 & 0 \\ 0 & 2 & 0 \\ 0 & 0 & 2 \end{pmatrix}}_{2I} + \begin{pmatrix} 0 & 2 & -1 \\ 0 & 0 & 2 \\ 0 & 0 & 0 \end{pmatrix}.$$

The sought matrix B is then:

$$B = \begin{pmatrix} 0 & 2 & -1 \\ 0 & 0 & 2 \\ 0 & 0 & 0 \end{pmatrix}.$$

(2) We have:

$$B^2 = \begin{pmatrix} 0 & 2 & -1 \\ 0 & 0 & 2 \\ 0 & 0 & 0 \end{pmatrix} \begin{pmatrix} 0 & 2 & -1 \\ 0 & 0 & 2 \\ 0 & 0 & 0 \end{pmatrix} = \begin{pmatrix} 0 & 0 & 4 \\ 0 & 0 & 0 \\ 0 & 0 & 0 \end{pmatrix}$$

and

$$B^3 = B^2 B = \begin{pmatrix} 0 & 0 & 4 \\ 0 & 0 & 0 \\ 0 & 0 & 0 \end{pmatrix} \begin{pmatrix} 0 & 2 & -1 \\ 0 & 0 & 2 \\ 0 & 0 & 0 \end{pmatrix} = \begin{pmatrix} 0 & 0 & 0 \\ 0 & 0 & 0 \\ 0 & 0 & 0 \end{pmatrix}.$$

(3) The best approach here is to apply the binomial theorem to $A = 2I + B$. We can do this because I and B commute. To proceed, we will need the powers of B. From the previous question, we know that:

$$B^n = \begin{pmatrix} 0 & 0 & 0 \\ 0 & 0 & 0 \\ 0 & 0 & 0 \end{pmatrix}, \quad \forall n \geq 3.$$

Applying the binomial expansion, we obtain:

$$A^n = (2I + B)^n = \sum_{p=0}^{n} \binom{n}{p} 2^{n-p} I^{n-p} B^p.$$

Since $I^{n-p} = I$, this simplifies to:

$$A^n = \sum_{p=0}^{n} \binom{n}{p} 2^{n-p} B^p.$$

Since $B^n = 0$ for $n \geq 3$, the sum reduces to:

$$A^n = \binom{n}{0} 2^n B^0 + \binom{n}{1} 2^{n-1} B^1 + \binom{n}{2} 2^{n-2} B^2 + 0 + \cdots + 0.$$

Explicitly, this gives:

$$A^n = 2^n I + n2^{n-1} B + \frac{n(n-1)}{2} 2^{n-2} B^2.$$

Rewriting the last term using $2^{n-2} = 2^{n-3} \cdot 2$, we obtain:

$$A^n = 2^n I + n2^{n-1} B + n(n-1)2^{n-3} B^2.$$

Finally, using the result from Question 1, we conclude:

$$A^n = \begin{pmatrix} 2^n & n2^n & n(n-2)2^{n-1} \\ 0 & 2^n & n2^n \\ 0 & 0 & 2^n \end{pmatrix}, \quad \forall n \in \mathbb{N}.$$

Exercise 1.3.15. *A matrix B is called a square root of a matrix A if $B^2 = A$.* Find two square roots of the matrix

$$A = \begin{pmatrix} 2 & 2 \\ 2 & 2 \end{pmatrix}.$$

Solution 1.3.15. Suppose that B is written as

$$B = \begin{pmatrix} a & b \\ c & d \end{pmatrix}.$$

Then,

$$B^2 = BB = \begin{pmatrix} a^2 + bc & ab + bd \\ ca + dc & bc + d^2 \end{pmatrix}.$$

To determine the square roots of A, we have to solve the following matrix equation:

$$\begin{pmatrix} a^2 + bc & ab + bd \\ ca + dc & bc + d^2 \end{pmatrix} = \begin{pmatrix} 2 & 2 \\ 2 & 2 \end{pmatrix}.$$

The reader can verify that the solutions of the preceding matrix equation are given by:

$$\begin{cases} a = 1, \\ b = 1, \\ c = 1, \\ d = 1, \end{cases} \quad \text{or} \quad \begin{cases} a = -1, \\ b = -1, \\ c = -1, \\ d = -1. \end{cases}$$

Thus, the two sought square roots are:

$$B_1 = \begin{pmatrix} 1 & 1 \\ 1 & 1 \end{pmatrix} \text{ and } B_2 = \begin{pmatrix} -1 & -1 \\ -1 & -1 \end{pmatrix}.$$

Exercise 1.3.16. (See also Exercise 7.3.18) Show that the matrix $A = \begin{pmatrix} 0 & 1 \\ 0 & 0 \end{pmatrix}$ does not have any square root.

Solution 1.3.16. Suppose A has a square root, noted B, which is a 2×2 matrix of the form

$$B = \begin{pmatrix} a & b \\ c & d \end{pmatrix}.$$

Then

$$B^2 = A \Longleftrightarrow \begin{pmatrix} a^2 + bc & ab + bd \\ ac + cd & bc + d^2 \end{pmatrix} = \begin{pmatrix} 0 & 1 \\ 0 & 0 \end{pmatrix}.$$

So, we need to solve the system

$$\begin{cases} a^2 + bc = 0, \\ ab + bd = 1, \\ ac + cd = 0, \\ bc + d^2 = 0. \end{cases}$$

The equation $(a + d)c = ac + cd = 0$ (commutativity is available here!) gives either $c = 0$ or $a = -d$. If $c = 0$, the first and the fourth equations of the system become $a^2 = 0$ and $d^2 = 0$, whereby $a = d = 0$. But, this is not consistent with the second equation (look at $0 = 1$!). Therefore, $c = 0$ does not yield anything.

Let us now examine the case $a = -d$. In this case, it is seen that

$$ab + bd = 1 \iff ab - ab = 1 \iff 0 = 1,$$

which is impossible.

Since both cases lead to contradictions, the system does not have any solutions. Therefore, the equation $B^2 = A$ is not satisfied by any matrix B, meaning that A does not have a square root.

Exercise 1.3.17. Show that the matrices $A = \begin{pmatrix} 1 & 0 \\ 0 & 0 \end{pmatrix}$ and $B = \begin{pmatrix} 1 & 1 \\ 0 & 1 \end{pmatrix}$ each have exactly two square roots.

Solution 1.3.17. Assume there is a matrix C of size 2×2 such that $C^2 = A$. If $C := \begin{pmatrix} a & b \\ c & d \end{pmatrix}$, then $C^2 = A$ is equivalent to

$$\begin{pmatrix} a^2 + bc & ab + bd \\ ca + dc & bc + d^2 \end{pmatrix} = \begin{pmatrix} 1 & 0 \\ 0 & 0 \end{pmatrix},$$

which then reduces to

$$\begin{cases} a^2 + bc = 1, \\ ab + bd = 0, \\ ac + cd = 0, \\ bc + d^2 = 0. \end{cases}$$

The third equation results in either $c = 0$ or $a = -d$. When $c = 0$, the fourth equation gives $d = 0$, and the first equation leads to $a = \pm 1$. Thus, the second equation yields $b = 0$.

Let us consider the scenario where $a = -d$. In this case, the second and third equations are both trivially satisfied. However, subtracting the fourth equation from the first equation, we get $a^2 - d^2 = 1$, which is inconsistent with $a = -d$ because we would end up with $0 = 1$, which is impossible. Thus, the given system possesses solutions only when $c = 0$, and so there are only two square roots of A, given by

$$\begin{pmatrix} 1 & 0 \\ 0 & 0 \end{pmatrix} \text{ and } \begin{pmatrix} -1 & 0 \\ 0 & 0 \end{pmatrix}.$$

A similar approach may be adopted to deal with the second matrix. Indeed, writing $C^2 = B$, where C is as above, yields

$$\begin{cases} a^2 + bc = 1, \\ ab + bd = 1, \\ ac + cd = 0, \\ bc + d^2 = 1. \end{cases}$$

Again, the third equation results in either $c = 0$ or $a = -d$. If $a = -d$, the second question is violated. So, $c = 0$ is the only possibility allowing further investigation. The first equation yields $a = \pm 1$ while the fourth equation leads to $d = \pm 1$. But, $a \neq -d$, and so $a = d = \pm 1$. The second equation now implies $b = 1/2a$. To summarize, matrix B has only two square roots given by

$$\begin{pmatrix} 1 & 1/2 \\ 0 & 1 \end{pmatrix} \text{ and } \begin{pmatrix} -1 & -1/2 \\ 0 & -1 \end{pmatrix}.$$

Exercise 1.3.18. Show that over $M_2(\mathbb{R})$, the identity matrix possesses infinitely many square roots.

Solution 1.3.18. Let A be a general 2×2 matrix:

$$A = \begin{pmatrix} a & b \\ c & d \end{pmatrix}.$$

Finding all square roots of I_2 means that we need to find all matrices that satisfy $A^2 = I_2$, that is,

$$A^2 = \begin{pmatrix} a^2 + bc & ab + bd \\ ac + cd & bc + d^2 \end{pmatrix} = \begin{pmatrix} 1 & 0 \\ 0 & 1 \end{pmatrix}.$$

Hence

$$\begin{cases} a^2 + bc = 1, \\ ab + bd = 0, \\ ac + cd = 0, \\ bc + d^2 = 1. \end{cases}$$

Solving the above system of equations allows us to find all possible square roots of the identity. Among the possible choices, we have the following family of square roots:

$$A_x = \begin{pmatrix} x & 1 \\ 1 - x^2 & -x \end{pmatrix}.$$

Then for each x,

$$A_x^2 = \begin{pmatrix} x & 1 \\ 1-x^2 & -x \end{pmatrix} \begin{pmatrix} x & 1 \\ 1-x^2 & -x \end{pmatrix}$$
$$= \begin{pmatrix} x^2 + 1 - x^2 & x - x \\ x(1-x^2) - x(1-x^2) & 1 - x^2 + x^2 \end{pmatrix}$$
$$= \begin{pmatrix} 1 & 0 \\ 0 & 1 \end{pmatrix}.$$

Another family is represented by $B_x = \begin{pmatrix} 1 & x \\ 0 & -1 \end{pmatrix}$, where $x \in \mathbb{R}$, and readers can readily verify that $B_x^2 = I_2$ for all $x \in \mathbb{R}$.

Exercise 1.3.19. Show that the number of nonzero square roots of a matrix is always even.

Solution 1.3.19. Suppose B is a nonzero square root of A, meaning that

$$B^2 = A.$$

Then the matrix $-B$ also satisfies

$$(-B)^2 = (-B)(-B) = B^2 = A.$$

Thus, whenever B is a square root of A, so is $-B$. This establishes that nonzero square roots of A always come in pairs of the form $\{B, -B\}$.

To conclude that the total number of nonzero square roots is even, we must ensure that every square root of A appears in such a pair. The only way a square root could be unpaired is if $B = -B$, which implies $B = 0$. Since we are only considering nonzero square roots, this case does not occur. Therefore, all nonzero square roots of A are accounted for in such pairs, proving that their total number is always even.

Exercise 1.3.20. Let A and B be two square matrices of the same order such that $AB = 0$, where 0 is the null matrix.
  (1) Show that if A is invertible, then $B = 0$.
  (2) Show that if B is invertible, then $A = 0$.
  (3) Deduce that if $AB = 0$, then either A is null, or B is null, or A and B are *both* non-invertible.

Solution 1.3.20.

  (1) We have:

$$AB = 0 \Longrightarrow \underbrace{A^{-1}A}_{I}B = A^{-1}0 = 0 \Longrightarrow B = 0.$$

(2) We have:

$$AB = 0 \Longrightarrow A \underbrace{BB^{-1}}_{I} = 0B^{-1} = 0 \Longrightarrow A = 0.$$

(3) According the above, if $AB = 0$, and $A \neq 0$, $B \neq 0$, then A and B are *both* non-invertible.

> **Exercise** 1.3.21. Let $A, B \in M_n$ be both invertible.
> (1) Show that A^{-1} is invertible and $(A^{-1})^{-1} = A$.
> (2) Show that AB is invertible and $(AB)^{-1} = B^{-1}A^{-1}$.
> (3) Show that $A + B$ is invertible if and only if $A^{-1} + B^{-1}$ is invertible.
>   **Hint:** Show that $A^{-1} + B^{-1} = A^{-1}(A + B)B^{-1}$.

Solution 1.3.21.

(1) Since A is invertible, we know that $AA^{-1} = I_n$. The latter equations also mean that A^{-1} is invertible and that $(A^{-1})^{-1} = A$, as suggested.

(2) Since A and B are invertible, we have $A^{-1}A = I_n$ and $B^{-1}B = I_n$. Thus, $A^{-1}AB = B$ and so $B^{-1}A^{-1}AB = B^{-1}B = I_n$. Since AB is a square matrix, we see that AB is indeed invertible with $(AB)^{-1} = B^{-1}A^{-1}$.

(3) First, we check the formula: $A^{-1} + B^{-1} = A^{-1}(A + B)B^{-1}$. We have

$$A^{-1}(A + B)B^{-1} = (A^{-1}A + A^{-1}B)B^{-1}$$
$$= (I + A^{-1}B)B^{-1}$$
$$= B^{-1} + A^{-1}BB^{-1}.$$

Thus, we obtain

$$A^{-1} + B^{-1} = A^{-1}(A + B)B^{-1}.$$

Assume now that $A + B$ is invertible. So, by the first question, $A^{-1}(A + B)$ is invertible as A^{-1} too is invertible. Since B^{-1} is also invertible, $A^{-1}(A + B)B^{-1}$ is invertible. To show the reverse implication, assume that $A^{-1} + B^{-1}$ is invertible. It is clear then that the required implication follows from the previous case by renaming A^{-1} and B^{-1} as A and B respectively.

Remark. Readers, and especially students, are invited to consult [5], where they can find an interesting discussion of when $(a + b)^{-1} = a^{-1} + b^{-1}$, where a and b are real, complex numbers or even matrices.

Exercise 1.3.22. Let A and B be two non-necessarily square matrices. If $AB = I$, then does it follow that $BA = I$?

Solution 1.3.22. The answer is negative. Given the finite-dimensional context we're in (whatever this means at present), a counter-example is only available if A and B are not square matrices. Consider

$$A = \begin{pmatrix} 1 & 0 & 0 \\ 0 & 1 & 0 \end{pmatrix} \text{ and } B = \begin{pmatrix} 1 & 0 \\ 0 & 1 \\ 0 & 0 \end{pmatrix}.$$

Then

$$AB = \begin{pmatrix} 1 & 0 \\ 0 & 1 \end{pmatrix} \text{ whereas } BA = \begin{pmatrix} 1 & 0 & 0 \\ 0 & 1 & 0 \\ 0 & 0 & 0 \end{pmatrix}.$$

In other words, $AB = I_2$ while $BA \neq I_3$.

Exercise 1.3.23. Show that the matrix $A = \begin{pmatrix} 1 & 0 & 0 \\ 0 & 1 & 0 \end{pmatrix}$ is not invertible.

Solution 1.3.23. In Exercise 1.3.22, we have already established that the matrix $\begin{pmatrix} 1 & 0 \\ 0 & 1 \\ 0 & 0 \end{pmatrix}$ serves as a right inverse of matrix A. For matrix A to be invertible, it would also need to have a left inverse, which would have to be of size 3×2. Let us assume that such a matrix exists and denote it as $B = \begin{pmatrix} a & b \\ c & d \\ e & f \end{pmatrix}$. We then have

$$\begin{pmatrix} 1 & 0 & 0 \\ 0 & 1 & 0 \\ 0 & 0 & 1 \end{pmatrix} = I_3 = \begin{pmatrix} a & b \\ c & d \\ e & f \end{pmatrix} \begin{pmatrix} 1 & 0 & 0 \\ 0 & 1 & 0 \end{pmatrix} = \begin{pmatrix} a & b & 0 \\ c & d & 0 \\ e & f & 0 \end{pmatrix}.$$

This leads to an impossible situation, specifically "$1 = 0$". In other words, the matrix B cannot exist, so A is non-invertible.

Exercise 1.3.24. Let $A, B, C \in M_n$ be such that $ABC = I$.
(1) Show that $BCA = I$.
(2) Is it true that $BAC = I$?

Solution 1.3.24.

(1) Suppose $ABC = I$. This implies that BC is invertible with an inverse given by A, so that $BCA = I$, as well.

(2) The answer, this time, is negative. Let

$$A = \begin{pmatrix} 1 & 0 \\ 0 & -1 \end{pmatrix}, \ B = \begin{pmatrix} 0 & 2 \\ 1 & 0 \end{pmatrix} \text{ and } C = \begin{pmatrix} 0 & -1 \\ 1/2 & 0 \end{pmatrix}.$$

Then $AB = \begin{pmatrix} 0 & 2 \\ -1 & 0 \end{pmatrix}$, and so $ABC = \begin{pmatrix} 1 & 0 \\ 0 & 1 \end{pmatrix} = I_2$.

Nonetheless, $BA = \begin{pmatrix} 0 & -2 \\ 1 & 0 \end{pmatrix}$, so $BAC = \begin{pmatrix} -1 & 0 \\ 0 & -1 \end{pmatrix} \neq I_2$, as needed.

Exercise 1.3.25. Let $A, B \in M_n$ satisfy the equation:

$$AB = A + B.$$

Prove that $AB = BA$.

Solution 1.3.25. Let I_n be the identity matrix. We have

$$(A - I_n)(B - I_n) = \underbrace{AB - A - B}_{=0} + I_n = I_n.$$

Thus, $A - I_n$ and $B - I_n$ are both invertible, and $(B - I_n)(A - I_n) = I_n$. Therefore,

$$(A - I_n)(B - I_n) = (B - I_n)(A - I_n),$$

which simplifies to:

$$AB = BA,$$

as required.

Exercise 1.3.26. Let $A = \begin{pmatrix} a & b \\ c & d \end{pmatrix}$, where a, b, c, and d are given scalars. Give a necessary and sufficient condition for A to be invertible, and find the expression of A^{-1}.

Solution 1.3.26. We claim that A is invertible if and only if $ad - bc \neq 0$ (the expression is usually called the determinant of A, noted $\det A$). To determine whether the *square* matrix A is invertible, we need to, e.g., solve the equation $AB = I_2$, where B is a 2×2 matrix, that is, we need to solve

$$\begin{pmatrix} a & b \\ c & d \end{pmatrix} \underbrace{\begin{pmatrix} x & y \\ z & t \end{pmatrix}}_{B} = \begin{pmatrix} 1 & 0 \\ 0 & 1 \end{pmatrix}.$$

The previous equation is equivalent to

$$\begin{cases} ax + bz = 1, \\ ay + bt = 0, \\ cx + dz = 0, \\ cy + dt = 1. \end{cases}$$

Multiplying the first equation by d and the third one by b implies that $adx + bdz = d$ and $bcx + bdz = 0$. This simplifies to $adx - bcx = 1$, or $(ad - bc)x = d$. If $\det A = ad - bc \neq 0$, then $x = d/\det A$. Similarly, we can solve for y, z, and t: $y = -b/\det A$, $z = -c/\det A$, and $t = a/\det A$. Thus, when $\det A \neq 0$, matrix A is invertible, and

$$A^{-1} = \frac{1}{\det A} \begin{pmatrix} d & -b \\ -c & a \end{pmatrix}.$$

Now, if $\det A = 0$, then A is not invertible. Indeed, if $ad = bc$, then $d = 0$ from the equation $(ad - bc)x = d$. The equation $(ad - bc)y = -b$ gives $b = 0$, and the remaining two equations lead to $c = 0$ and $a = 0$. Thus, $A = 0$, which violates the equation $AB = I_2$. Accordingly, A is not invertible.

Exercise 1.3.27. On M_n, define a binary relation $\mathcal{R}$ by

$$A\mathcal{R}B \iff \exists \text{ an invertible } T \in M_n : T^{-1}AT = B.$$

Show that $\mathcal{R}$ is an equivalence relation on M_n.

Solution 1.3.27. To see why $\mathcal{R}$ is reflexive, let $A \in M_n$. If $T = I$, the identity matrix, which is invertible, then $I^{-1}AI = A$, indicating that $\mathcal{R}$ is reflexive.

Next, let $A, B \in M_n$ be such that $A\mathcal{R}B$, that is, $T^{-1}AT = B$ for some invertible $T \in M_n$. Since T^{-1} is invertible and $A = (T^{-1})^{-1}BT^{-1}$, $B\mathcal{R}A$, which indicates that $\mathcal{R}$ is symmetric.

Finally, let $A, B, C \in M_n$ be such that $A\mathcal{R}B$ and $B\mathcal{R}C$, signifying that $T^{-1}AT = B$ and $S^{-1}BS = C$ for certain invertible $T, S \in M_n$. Then

$$(TS)^{-1}ATS = S^{-1}T^{-1}ATS = S^{-1}BS = C,$$

meaning that $A\mathcal{R}C$ as TS is invertible. Thus, $\mathcal{R}$ is an equivalence relation on M_n, as wished.

Exercise 1.3.28. Let $A, B \in M_n$ be such that $A^2 = 0$, while $B^2 \neq 0$. Can A be similar to B?

Solution 1.3.28. The answer is no! Assume that A and B are similar, meaning that there exists an invertible matrix $T \in M_n$ such that $T^{-1}AT = B$. Then, we can compute B^2 as follows:

$$B^2 = (T^{-1}AT)^2 = T^{-1}ATT^{-1}AT = T^{-1}A^2T.$$

Now, since $A^2 = 0$ by assumption, we have:

$$T^{-1}A^2T = T^{-1}0T = 0.$$

Thus, $B^2 = 0$, which contradicts the assumption that $B^2 \neq 0$. Therefore, the assumption that A and B are similar leads to a contradiction, and the answer is indeed no.

Exercise 1.3.29. Let $A, B \in M_n$.

(1) Is AB always similar to BA?

(2) Show that if, for instance, A is invertible, then AB and BA are similar.

Solution 1.3.29.

(1) The answer is negative. For example, take $A = \begin{pmatrix} 0 & 0 \\ 0 & 1 \end{pmatrix}$ and $B = \begin{pmatrix} 0 & 1 \\ 0 & 0 \end{pmatrix}$. Then, $AB = 0$, whereas $BA = B$. In other words, AB is not similar to BA.

(2) Suppose A is invertible without loss of generality. Writing

$$BA = A^{-1}ABA = A^{-1}(AB)A,$$

shows that AB and BA are similar.

Exercise 1.3.30. Let $A, B \in M_n(\mathbb{R})$.

(1) If $A^2 = B^2$, then does it follow that $A = B$?

(2) Let p and q be two relatively prime numbers such that $A^p = B^p$ and $A^q = B^q$. Show that $A = B$ whenever A is invertible.

Solution 1.3.30.

(1) The answer is no. For instance, let $A = \begin{pmatrix} 1 & 0 \\ 0 & -1 \end{pmatrix}$. Then $A \neq I_2$, and yet

$$A^2 = \begin{pmatrix} 1 & 0 \\ 0 & 1 \end{pmatrix} = I_2^2.$$

(2) Since A is invertible, it is seen that B too is invertible. Reason: By the invertibility of A, we get that of A^p or that of B^p. So, e.g. for some $C \in M_n$, $B^pC = I_n$ and hence $B(B^{p-1}C) = I$, whereby B is invertible.

Since p and q are relatively prime numbers, Bézout's theorem in arithmetic says that $up + vq = 1$ for some integers u and v (only one of them is negative). WLOG, suppose that u is the negative integer. Now, $A^p = B^p$ yields $A^{up} = B^{up}$, and $A^q = B^q$ implies that $A^{vq} = B^{vq}$. Therefore, $A^{up}A^{vq} = B^{up}B^{vq}$, and so

$$A = A^{up+vq} = B^{up+vq} = B,$$

as wished.

Remark. We saw above that $A^2 = I$ does not give $A = I$ as the only possibility. But, and as an application of the foregoing question, if $A^2 = I$ ($= I^2$) and $A^3 = I$, then the only outcome is $A = I$ because 2 and 3 are co-prime. Observe in the end that the invertibility of A was tacitly assumed in the condition $A^2 = I$.

Exercise 1.3.31. Let $A \in M_n$ be nilpotent of index p. Show that $I - A$ is invertible with
$$(I - A)^{-1} = I + A + A^2 + \cdots + A^{p-1}.$$

Solution 1.3.31. By assumption, $A^p = 0$. Since

$$(I - A)(I + A + A^2 + \cdots + A^{p-1})$$
$$= I + A + A^2 + \cdots + A^{p-1} - A - A^2 + \cdots - A^{p-1} - A^p$$
$$= I,$$

we see that $I - A$ is invertible and
$$(I - A)^{-1} = I + A + A^2 + \cdots + A^{p-1}.$$

Exercise 1.3.32. Let A and B be two square matrices of order n. Show that, if one of the two matrices is invertible, then AB and BA are similar.

Solution 1.3.32. WLOG, assume A is invertible. Then

$$A^{-1}(AB)A = BA,$$

which indicates that the matrices AB and BA are similar.

Exercise 1.3.33. Define the binary relation $\mathcal{R}$ on the set of matrices $n \times n$ by:
$$A\mathcal{R}B \iff AB = BA.$$
Is $\mathcal{R}$ reflexive? Is it symmetric? Is it transitive?

Solution 1.3.33. Clearly the relation $\mathcal{R}$ is reflexive (because every matrix commutes with itself). It is also obvious that $\mathcal{R}$ is symmetric.

However, $\mathcal{R}$ is not transitive. Let A and C be any two *non-commuting* matrices (e.g. $A = \begin{pmatrix} 0 & 1 \\ 1 & 0 \end{pmatrix}$ and $C = \begin{pmatrix} 2 & 0 \\ 0 & 1 \end{pmatrix}$). By setting, $B = 0$, we see that A commutes with B and B commutes with C, but A does not commute with C.

Since B above is not invertible, one could think that perhaps the relation is transitive on the set of invertible matrices. This is still not sufficient, as may be seen by letting $B = I$ and taking A and C as above (which are two *non-commuting invertible* matrices).

Exercise 1.3.34. Consider the following matrix:

$$A = \begin{pmatrix} 1 & 0 & 1 \\ 1 & 1 & 0 \\ 0 & 1 & 1 \end{pmatrix}.$$

Show that A is invertible and find its inverse A^{-1}.

Solution 1.3.34. Since A is a square matrix of order 3, A is invertible if there exists a matrix B, of order 3, such that

$$AB = I = \begin{pmatrix} 1 & 0 & 0 \\ 0 & 1 & 0 \\ 0 & 0 & 1 \end{pmatrix}.$$

We suppose that B is of the form:

$$B = \begin{pmatrix} a & b & c \\ d & e & f \\ g & h & i \end{pmatrix},$$

and we have to determine its entries. We have:

$$AB = I \Longleftrightarrow \begin{pmatrix} 1 & 0 & 1 \\ 1 & 1 & 0 \\ 0 & 1 & 1 \end{pmatrix} \begin{pmatrix} a & b & c \\ d & e & f \\ g & h & i \end{pmatrix} = \begin{pmatrix} 1 & 0 & 0 \\ 0 & 1 & 0 \\ 0 & 0 & 1 \end{pmatrix}$$

$$\Longleftrightarrow \begin{pmatrix} a+g & b+h & c+i \\ a+d & b+e & c+f \\ d+g & e+h & f+i \end{pmatrix} = \begin{pmatrix} 1 & 0 & 0 \\ 0 & 1 & 0 \\ 0 & 0 & 1 \end{pmatrix}$$

$$\Longleftrightarrow \begin{cases} a+g = 1 \\ b+h = 0 \\ c+i = 0 \\ a+d = 0 \\ b+e = 1 \\ c+f = 0 \\ d+g = 0 \\ e+h = 0 \\ f+i = 1 \end{cases}$$

Fortunately, in this case, we can solve "$a+g = 1$, $a+d = d+g = 0$", "$b + h = e + h = 0$, $b + e = 1$", and "$c + i = c + f = 0$, $f + i = 1$" separately one from another. The first system gives: $a = g = \frac{1}{2}$, $d = -\frac{1}{2}$; $b = e = \frac{1}{2}$, $h = -\frac{1}{2}$; $c = -\frac{1}{2}$, $f = i = \frac{1}{2}$. Hence,

$$A^{-1} = B = \begin{pmatrix} \frac{1}{2} & \frac{1}{2} & -\frac{1}{2} \\ -\frac{1}{2} & \frac{1}{2} & \frac{1}{2} \\ \frac{1}{2} & -\frac{1}{2} & \frac{1}{2} \end{pmatrix}.$$

Remark. This method is not very practical; it becomes even more tedious when the matrix is of order greater than 3 or contains large numbers. In fact, this is a method one might only need to use once in a lifetime.

Exercise 1.3.35. Let $0 < \theta < \frac{\pi}{2}$ and $x \in \mathbb{R}$. Find the inverse of the following matrices:

$$A = \begin{pmatrix} \cos\theta & \sin\theta \\ -\sin\theta & \cos\theta \end{pmatrix} \text{ and } B = \begin{pmatrix} \frac{1}{2}(e^x + e^{-x}) & \frac{1}{2}(e^x - e^{-x}) \\ \frac{1}{2}(e^x - e^{-x}) & \frac{1}{2}(e^x + e^{-x}) \end{pmatrix}.$$

Solution 1.3.35. Let us determine the matrix X such that

$$AX = I_2 = \begin{pmatrix} 1 & 0 \\ 0 & 1 \end{pmatrix}.$$

Supposing that

$$X = \begin{pmatrix} a & b \\ c & d \end{pmatrix},$$

we obtain the equation

$$AX = I_2 \iff \begin{pmatrix} \cos\theta & \sin\theta \\ -\sin\theta & \cos\theta \end{pmatrix} \begin{pmatrix} a & b \\ c & d \end{pmatrix} = \begin{pmatrix} 1 & 0 \\ 0 & 1 \end{pmatrix}.$$

Expanding, we obtain the following system:

$$\begin{cases} a\cos\theta + c\sin\theta = 1, \\ b\cos\theta + d\sin\theta = 0, \\ -a\sin\theta + c\cos\theta = 0, \\ -b\sin\theta + d\cos\theta = 1. \end{cases}$$

Multiplying the first equation by $\sin\theta$ and the third equation by $\cos\theta$, then summing, we find:

$$c\underbrace{(\sin^2\theta + \cos^2\theta)}_{=1} = \sin\theta \implies c = \sin\theta.$$

From the third equation, we obtain a:

$$a\sin\theta = \sin\theta\cos\theta \implies a = \cos\theta, \quad (\text{since } \sin\theta \neq 0).$$

Similarly, we obtain:

$$b = -\sin\theta, \quad d = \cos\theta.$$

Thus, the inverse matrix is given by:

$$A^{-1} = \begin{pmatrix} \cos\theta & -\sin\theta \\ \sin\theta & \cos\theta \end{pmatrix}.$$

To find the inverse of matrix B, recall that:

$$\cosh x = \frac{1}{2}(e^x + e^{-x}), \quad \sinh x = \frac{1}{2}(e^x - e^{-x}),$$

and the identity:

$$\cosh^2 x - \sinh^2 x = 1, \quad \forall x \in \mathbb{R}.$$

We need to invert the matrix:

$$B = \begin{pmatrix} \frac{1}{2}(e^x + e^{-x}) & \frac{1}{2}(e^x - e^{-x}) \\ \frac{1}{2}(e^x - e^{-x}) & \frac{1}{2}(e^x + e^{-x}) \end{pmatrix} = \begin{pmatrix} \cosh x & \sinh x \\ \sinh x & \cosh x \end{pmatrix}.$$

If $Y = \begin{pmatrix} a & b \\ c & d \end{pmatrix}$ is its inverse, then after a few calculations, we obtain the following system to solve:

$$\begin{cases} a \cosh x + b \sinh x = 1, \\ a \sinh x + b \cosh x = 0, \\ c \cosh x + d \sinh x = 0, \\ c \sinh x + d \cosh x = 1. \end{cases}$$

Multiplying the first equation by $\sinh x$ and the second one by $\cosh x$, then subtracting, we find:

$$a \underbrace{(\cosh^2 x - \sinh^2 x)}_{=1} = \cosh x \implies a = \cosh x.$$

Thus,

$$\cosh x \sinh x + b \cosh x = 0 \implies b = -\sinh x.$$

Similarly, we obtain:

$$c = -\sinh x, \quad d = \cosh x.$$

Finally, the inverse of B is:

$$B^{-1} = \begin{pmatrix} \cosh x & -\sinh x \\ -\sinh x & \cosh x \end{pmatrix} = \begin{pmatrix} \frac{1}{2}(e^x + e^{-x}) & -\frac{1}{2}(e^x - e^{-x}) \\ -\frac{1}{2}(e^x - e^{-x}) & \frac{1}{2}(e^x + e^{-x}) \end{pmatrix}.$$

Exercise 1.3.36. Let

$$A = \begin{pmatrix} 1 & 4 \\ 0 & -1 \end{pmatrix}.$$

By calculating A^2, show that A is invertible and deduce the inverse matrix A^{-1}.

Solution 1.3.36. We have

$$A^2 = AA = \begin{pmatrix} 1 & 4 \\ 0 & -1 \end{pmatrix}\begin{pmatrix} 1 & 4 \\ 0 & -1 \end{pmatrix} = \begin{pmatrix} 1 & 0 \\ 0 & 1 \end{pmatrix} = I_2.$$

Since $A^2 = I_2$, we get $AA = I_2$, which clearly shows that A is invertible, and its inverse is itself, i.e.,

$$A^{-1} = A = \begin{pmatrix} 1 & 4 \\ 0 & -1 \end{pmatrix}.$$

Exercise 1.3.37. Let A be the matrix:

$$\begin{pmatrix} -1 & 1 & 1 \\ 1 & -1 & 1 \\ 1 & 1 & -1 \end{pmatrix}.$$

(1) Check that $A^2 = 2I - A$, where I is the identity matrix of $M_3(\mathbb{R})$.
(2) Deduce that A is invertible and give A^{-1}.

Solution 1.3.37.

(1) The verification of this formula is left to the reader.
(2) We have:

$$A^2 = 2I - A \iff A^2 + A = 2I \iff \frac{1}{2}(A + I)A = I.$$

This shows that A is invertible, and its inverse is given by $\frac{1}{2}(A + I)$.

Exercise 1.3.38. Suppose that the matrix A satisfies

$$(I + 2A)^{-1} = \begin{pmatrix} -1 & 2 \\ 4 & 5 \end{pmatrix}.$$

Find A.

Solution 1.3.38. It is the inverse process! Set $B = (I + 2A)^{-1}$. Then, by the definition of an inverse matrix, we have:

$$B(I + 2A) = I_2.$$

Assume that

$$A = \begin{pmatrix} a & b \\ c & d \end{pmatrix}, \quad \text{hence} \quad I + 2A = \begin{pmatrix} 2a + 1 & 2b \\ 2c & 2d + 1 \end{pmatrix}.$$

Let us determine the values of a, b, c, and d satisfying the equation:

$$\begin{pmatrix} -1 & 2 \\ 4 & 5 \end{pmatrix} \begin{pmatrix} 2a + 1 & 2b \\ 2c & 2d + 1 \end{pmatrix} = \begin{pmatrix} 1 & 0 \\ 0 & 1 \end{pmatrix}.$$

Expanding the left-hand side, we obtain:

$$\begin{pmatrix} -2a - 1 + 4c & -2b + 4d + 2 \\ 8a + 4 + 10c & 8b + 10d + 5 \end{pmatrix} = \begin{pmatrix} 1 & 0 \\ 0 & 1 \end{pmatrix}.$$

Solving the system (left to the reader), we find:

$$\begin{cases} a = -\frac{9}{13}, \\ b = \frac{1}{13}, \\ c = \frac{2}{13}, \\ d = -\frac{6}{13}. \end{cases}$$

Thus, the sought matrix is:

$$A = \begin{pmatrix} -\frac{9}{13} & \frac{1}{13} \\ \frac{2}{13} & -\frac{6}{13} \end{pmatrix}.$$

Exercise 1.3.39. Using the method of Gaussian elimination find the inverse of the matrix:

$$A = \begin{pmatrix} 1 & 2 & 3 \\ 2 & 5 & 3 \\ 1 & 0 & 8 \end{pmatrix}.$$

Solution 1.3.39. The calculation goes as follows:

$$\left(\begin{array}{ccc|ccc} 1 & 2 & 3 & 1 & 0 & 0 \\ 2 & 5 & 3 & 0 & 1 & 0 \\ 1 & 0 & 8 & 0 & 0 & 1 \end{array} \right) \xrightarrow{(-2R_1+R_2 \to R_2, \ -R_1+R_3 \to R_3)}$$

$$\left(\begin{array}{ccc|ccc} 1 & 2 & 3 & 1 & 0 & 0 \\ 0 & 1 & -3 & -2 & 1 & 0 \\ 0 & -2 & 5 & -1 & 0 & 1 \end{array} \right) \xrightarrow{(2R_2+R_3 \to R_3)}$$

$$\left(\begin{array}{ccc|ccc} 1 & 2 & 3 & 1 & 0 & 0 \\ 0 & 1 & -3 & -2 & 1 & 0 \\ 0 & 0 & -1 & -5 & 2 & 1 \end{array} \right) \xrightarrow{(-R_3 \to R_3)}$$

$$\left(\begin{array}{ccc|ccc} 1 & 2 & 3 & 1 & 0 & 0 \\ 0 & 1 & -3 & -2 & 1 & 0 \\ 0 & 0 & 1 & 5 & -2 & -1 \end{array} \right) \xrightarrow{(3R_3+R_2 \to R_2, \ -3R_3+R_1 \to R_1)}$$

$$\left(\begin{array}{ccc|ccc} 1 & 2 & 0 & -14 & 6 & 3 \\ 0 & 1 & 0 & 13 & -5 & -3 \\ 0 & 0 & 1 & 5 & -2 & -1 \end{array} \right) \xrightarrow{(-2R_2+R_1 \to R_1)}$$

$$\left(\begin{array}{ccc|ccc} 1 & 0 & 0 & -40 & 16 & 9 \\ 0 & 1 & 0 & 13 & -5 & -3 \\ 0 & 0 & 1 & 5 & -2 & -1 \end{array} \right).$$

Hence:

$$A^{-1} = \begin{pmatrix} -40 & 16 & 9 \\ 13 & -5 & -3 \\ 5 & -2 & -1 \end{pmatrix}.$$

Exercise 1.3.40. Using the method of Gaussian elimination find the inverse of the matrix:

$$A = \begin{pmatrix} 0 & 1 & -1 \\ 4 & -3 & 4 \\ 3 & -3 & 4 \end{pmatrix}.$$

Solution 1.3.40.

$$\left(\begin{array}{ccc|ccc} 0 & 1 & -1 & 1 & 0 & 0 \\ 4 & -3 & 4 & 0 & 1 & 0 \\ 3 & -3 & 4 & 0 & 0 & 1 \end{array} \right)$$

$$\left(\begin{array}{ccc|ccc} 4 & -3 & 4 & 0 & 1 & 0 \\ 0 & 1 & -1 & 1 & 0 & 0 \\ 3 & -3 & 4 & 0 & 0 & 1 \end{array} \right) \quad \text{(after } R_1 \leftrightarrow R_2\text{)}.$$

$$\left(\begin{array}{ccc|ccc} 4 & -3 & 4 & 0 & 1 & 0 \\ 0 & 1 & -1 & 1 & 0 & 0 \\ 0 & -3 & 4 & 0 & -3 & 4 \end{array} \right) \quad \text{(after } -3R_1 + 4R_3 \to R_3\text{)}.$$

$$\left(\begin{array}{ccc|ccc} 4 & 0 & 1 & 3 & 1 & 0 \\ 0 & 1 & -1 & 1 & 0 & 0 \\ 0 & 0 & 1 & 3 & -3 & 4 \end{array} \right) \quad (R_1 + 3R_2 \to R_1 \text{ and } 3R_2 + R_3 \to R_3).$$

$$\left(\begin{array}{ccc|ccc} 4 & 0 & 0 & 0 & 4 & -4 \\ 0 & 1 & 0 & 4 & -3 & 4 \\ 0 & 0 & 1 & 3 & -3 & 4 \end{array} \right) \quad \text{(after } R_1 - R_3 \to R_1 \text{ and } R_2 + R_3 \to R_2\text{)}.$$

Finally:

$$\left(\begin{array}{ccc|ccc} 1 & 0 & 0 & 0 & 1 & -1 \\ 0 & 1 & 0 & 4 & -3 & 4 \\ 0 & 0 & 1 & 3 & -3 & 4 \end{array} \right) \quad \text{(after } \frac{1}{4}R_1 \to R_1\text{)}.$$

Whence:

$$A^{-1} = \begin{pmatrix} 0 & 1 & -1 \\ 4 & -3 & 4 \\ 3 & -3 & 4 \end{pmatrix}$$

(we notice that $A^{-1} = A$).

Exercise 1.3.41. Using the method of Gaussian elimination, show that the matrix:

$$A = \begin{pmatrix} 1 & 6 & 4 \\ 2 & 4 & -1 \\ -1 & 2 & 5 \end{pmatrix}$$

is not invertible.

Solution 1.3.41. We have:

$$\left(\begin{array}{ccc|ccc} 1 & 6 & 4 & 1 & 0 & 0 \\ 2 & 4 & -1 & 0 & 1 & 0 \\ -1 & 2 & 5 & 0 & 0 & 1 \end{array} \right)$$

$$\left(\begin{array}{ccc|ccc} 1 & 6 & 4 & 1 & 0 & 0 \\ 0 & -8 & -9 & -2 & 1 & 0 \\ 0 & 8 & 9 & 1 & 0 & 1 \end{array} \right) \quad (-2R_1 + R_2 \to R_2 \text{ and } R_1 + R_3 \to R_3).$$

$$\left(\begin{array}{ccc|ccc} 1 & 6 & 4 & 1 & 0 & 0 \\ 0 & -8 & -9 & -2 & 1 & 0 \\ 0 & 0 & 0 & -1 & 1 & 1 \end{array} \right) \quad (\text{after } R_2 + R_3 \to R_3).$$

Since we have obtained a row of zeros on the left side, we conclude that A is not invertible.

Exercise 1.3.42. Compute $\mathrm{tr}A$ when possible:

$$1)\ A = \begin{pmatrix} -1 & -50000 & 100 \\ 2011 & 0 & -98478 \\ 0 & -458 & 5 \end{pmatrix}, \quad 2)\ A = \begin{pmatrix} 0 & 0 \\ 256 & 1 \\ 12 & -98 \end{pmatrix},$$

$$3)\ A = \begin{pmatrix} -2 & 0 \\ 0 & 2 \end{pmatrix}.$$

Solution 1.3.42. In the first case, we get:

$$\mathrm{tr}A = -1 + 0 + 5 = 4.$$

In the second case, $\mathrm{tr}A$ is not well-defined because A is not a square matrix.

In the third case, we find:

$$\mathrm{tr}A = -2 + 2 = 0.$$

Exercise 1.3.43. Let A be a complex square matrix of order n, i.e.,

$$A = \begin{pmatrix} a_{11} & a_{12} & \cdots & a_{1n} \\ a_{21} & a_{22} & \cdots & a_{2n} \\ \cdots & \cdots & \cdots & \cdots \\ a_{n1} & a_{n2} & \cdots & a_{nn} \end{pmatrix},$$

where entries are possibly complex numbers. Set

$$\overline{A} = \begin{pmatrix} \overline{a_{11}} & \overline{a_{12}} & \cdots & \overline{a_{1n}} \\ \overline{a_{21}} & \overline{a_{22}} & \cdots & \overline{a_{2n}} \\ \cdots & \cdots & \cdots & \cdots \\ \overline{a_{n1}} & \overline{a_{n2}} & \cdots & \overline{a_{nn}} \end{pmatrix},$$

where $\overline{a_{ij}}$ is the complex conjugate of a_{ij}. Check that

$$\text{tr}(\overline{A}) = \overline{\text{tr}(A)}.$$

Solution 1.3.43. Since $\text{tr}(A) = a_{11} + a_{22} + \cdots + a_{nn}$, it follows that

$$\overline{\text{tr}(A)} = \overline{a_{11} + a_{22} + \cdots + a_{nn}} = \overline{a_{11}} + \overline{a_{22}} + \cdots + \overline{a_{nn}} = \text{tr}(\overline{A}),$$

as suggested.

Exercise 1.3.44. Let A be a square matrix of order n. Let $\text{tr}A$ be its trace, which is a function from $M_n(\mathbb{R})$ into $\mathbb{R}$. Is it surjective? Is it injective?

Solution 1.3.44. Function "tr" is surjective if and only if

$$\forall y \in \mathbb{R} : \exists A \in M_n(\mathbb{R}) : \text{tr}A = y.$$

Let $y \in \mathbb{R}$. Consider the following square matrix of order n:

$$A = \begin{pmatrix} y & 0 & \cdots & 0 \\ 0 & 0 & \ddots & \vdots \\ \vdots & \ddots & \ddots & 0 \\ 0 & \cdots & 0 & 0 \end{pmatrix}.$$

Then, $\text{tr}A = y + 0 + 0 \cdots + 0 = y$, meaning that "tr" is surjective.

Now, "tr" is not injective. To do this, we need to find two matrices A and B such that $A \neq B$, yet $\text{tr}A = \text{tr}B$. Consider the following square matrices of order n:

$$A = \begin{pmatrix} -1 & 0 & 0 & \cdots & 0 \\ 0 & 1 & 0 & \cdots & 0 \\ 0 & 0 & 0 & \cdots & 0 \\ \vdots & \vdots & \vdots & \ddots & \vdots \\ 0 & \cdots & 0 & 0 & 0 \end{pmatrix} \quad \text{and} \quad B = \begin{pmatrix} 1 & 0 & 0 & \cdots & 0 \\ 0 & -1 & 0 & \cdots & 0 \\ 0 & 0 & 0 & \cdots & 0 \\ \vdots & \vdots & \vdots & \ddots & \vdots \\ 0 & \cdots & 0 & 0 & 0 \end{pmatrix}.$$

We observe that $A \neq B$, yet

$$\mathrm{tr}A = -1 + 1 = 0 = 1 - 1 = \mathrm{tr}B,$$

showing that "tr" is not one-to-one.

Exercise 1.3.45. Prove the nonexistence of square matrices of order n (with $n \geq 1$), A and B such that

$$AB - BA = I_n,$$

where I is the identity matrix of order n.

Solution 1.3.45. Assume, for the sake of contradiction, that such matrices A and B exist, that is, $AB - BA = I_n$. Then,

$$\mathrm{tr}(AB - BA) = \mathrm{tr}I_n = 1 + 1 + \cdots + 1 = n,$$

and since $\mathrm{tr}(AB - BA) = \mathrm{tr}(AB) - \mathrm{tr}(BA) = 0$, we find $0 = n$, which is absurd! Thus, no such matrices exist.

Exercise 1.3.46. Let $A, B \in M_n(\mathbb{R})$.
(1) Assume that $\mathrm{tr}(A^tA) = 0$. Show that $A = 0$.
(2) Infer that each of the equations $A^tA = 0$ or $AA^t = 0$ implies $A = 0$.
(3) If $AA^tA = 0$, does it follow that $A = 0$?
(4) Suppose $\mathrm{tr}(XA) = \mathrm{tr}(XB)$ for all $X \in M_n(\mathbb{R})$. Infer that $A = B$.

Solution 1.3.46.
(1) If $A = (a_{ij})$, where $1 \leq i, j \leq n$, then the diagonal elements of A^tA are of the form

$$\alpha_{ii} = \sum_{k=1}^{n} a_{ik}^2.$$

Since $\mathrm{tr}(A^tA) = 0$, we must have $\sum_{i=1}^{n} \alpha_{ii} = 0$. Since each α_{ii} is nonnegative, we have $\alpha_{11} = \alpha_{22} = \cdots = \alpha_{nn} = 0$. Thus, $\sum_{k=1}^{n} a_{ik}^2 = 0$ for all i, which due to the nonnegativity of a_{ik}^2, then implies that $a_{ij} = 0$ for all $1 \leq i, j \leq n$. Consequently, $A = 0$, as required.

Remark. The result of the previous question also applies to a non-square matrix. Interested readers can adapt the proof provided previously to this case. Without further notice, we will use this result for both square and non-square matrices.

Remark. If A has complex entries, then $\mathrm{tr}(A^tA) = 0$ does not necessarily yield $A = 0$. A counterexample is $A =$

$\begin{pmatrix} 1 & i \\ -i & 1 \end{pmatrix}$, where it can be shown that $\text{tr}(A^tA) = 0$, yet $A \neq 0$.

See Exercise 5.4.8 for a similar result about the complex case.

If $A \in M_n(\mathbb{F})$, where $\mathbb{F}$ is a field that differs from $\mathbb{R}$ and $\mathbb{C}$, the result may still not hold. For instance, if $A = \begin{pmatrix} 1 & 1 \\ 1 & 1 \end{pmatrix}$ is considered as a matrix with entries in $\mathbb{Z}_2$, then $A \neq 0$, yet

$$A^tA = \begin{pmatrix} 1 & 1 \\ 1 & 1 \end{pmatrix}\begin{pmatrix} 1 & 1 \\ 1 & 1 \end{pmatrix} = \begin{pmatrix} 0 & 0 \\ 0 & 0 \end{pmatrix}$$

because, in $\mathbb{Z}_2$, $1 + 1 = 0$.

(2) If $A^tA = 0$ (resp. $AA^t = 0$), then $\text{tr}(A^tA) = 0$ (resp. $\text{tr}(AA^t) = 0$). The preceding question implies $A = 0$ in the former case, and $A^t = 0$ in the latter case, whereby $A = 0$ in either case.

(3) The answer is positive. Since $AA^tA = 0$, it follows that $A^tAA^tA = 0$, or equivalently, $(A^tA)^tA^tA = 0$. Thus, $B^tB = 0$, where we have set $B = A^tA$. By the previous question, $B = 0$, i.e., $A^tA = 0$, which leads to $A = 0$, as desired.

Remark. It is clear that $A^tAA^t = 0$ implies $A = 0$.

(4) By assumption, $\text{tr}(XA) = \text{tr}(XB)$ holds for *all* X. Hence,

$$0 = \text{tr}(XA) - \text{tr}(XB) = \text{tr}(XA - XB) = \text{tr}[X(A - B)],$$

still for all X. By choosing $X = (A - B)^t$ and using the result of the preceding question, we conclude that $A - B = 0$, or $A = B$, as suggested.

1.4. Exercises without Solutions

Exercise 1.4.1. Assume that A, B, C and D are matrices of sizes:

$$(4 \times 5), \ (4 \times 5), \ (5 \times 2) \text{ and } (4 \times 2).$$

respectively. Determine the size of the following matrices (in the case where they are well-defined):

$$BA, \ AC + D, \ AB + B, A^t + A, A^tC, \ A + D?$$

Exercise 1.4.2. Is there a square matrix A of order 2 such that

$$A\begin{pmatrix} a & b \\ c & d \end{pmatrix} = \begin{pmatrix} b & d \\ a & c \end{pmatrix}?$$

Exercise 1.4.3. Let $\alpha \in \mathbb{R}$ and let

$$A = \begin{pmatrix} \cos\alpha & -\sin\alpha \\ \sin\alpha & \cos\alpha \end{pmatrix}.$$

(1) Show that:

$$A^2 = \begin{pmatrix} \cos 2\alpha & -\sin 2\alpha \\ \sin 2\alpha & \cos 2\alpha \end{pmatrix} \text{ and } A^3 = \begin{pmatrix} \cos 3\alpha & -\sin 3\alpha \\ \sin 3\alpha & \cos 3\alpha \end{pmatrix}.$$

(2) Find the general formula of A^n (where $n \in \mathbb{N}$) and prove it by induction.

Exercise 1.4.4. Let $n \in \mathbb{N}$. Determine A^n where A is the matrix:

$$A = \begin{pmatrix} x & 0 & 0 \\ 0 & y & 0 \\ 1 & 0 & z \end{pmatrix},$$

where x, y, and z are real scalars. **Hint:** Use the formula:

$$\forall a \neq b: \ a^n + a^{n-1}b + a^{n-2}b^2 + \cdots + ab^{n-1} + b^n = \frac{a^{n+1} - b^{n+1}}{a - b}.$$

Exercise 1.4.5. Let A be the matrix:

$$A = \begin{pmatrix} 2 & 0 \\ 4 & 1 \end{pmatrix}.$$

Find A^3, A^{-3}, and $A^2 - 2A + I$.

Exercise 1.4.6. Let A and B be square matrices that satisfy $A^2 = B^2 = (BA)^2 = I$. Show that $AB = BA$.

Exercise 1.4.7. Find, using the method of Gaussian elimination, the inverse of each of the following matrices:

$$A = \begin{pmatrix} 1 & 4 \\ 2 & 7 \end{pmatrix}, \ B = \begin{pmatrix} 1 & 0 & 1 \\ 0 & 1 & 1 \\ 1 & 1 & 0 \end{pmatrix}, \ C = \begin{pmatrix} 0 & 0 & 2 & 0 \\ 1 & 0 & 0 & 1 \\ 0 & -1 & 3 & 0 \\ 2 & 1 & 5 & -3 \end{pmatrix}.$$

Exercise 1.4.8. Prove that the matrix

$$A = \begin{pmatrix} 0 & a & 0 & 0 & 0 \\ b & 0 & c & 0 & 0 \\ 0 & d & 0 & e & 0 \\ 0 & 0 & f & 0 & g \\ 0 & 0 & 0 & h & 0 \end{pmatrix}$$

is not invertible, regardless of the values of its entries.

Exercise 1.4.9. Show that an upper triangular matrix is similar to a lower triangular matrix.

Exercise 1.4.10. Let
$$A = \begin{pmatrix} 2 & 5 & -3 \\ 2 & 1 & 1 \\ 2 & 0 & -1 \end{pmatrix}.$$
(1) Prove that A satisfies: $A^3 - A^2 - 5A - 24I = 0$.
(2) What is the inverse of A?

Exercise 1.4.11. Let I be the identity matrix of $M_3(\mathbb{R})$ and let:
$$A = \begin{pmatrix} 1 & 1 & 0 \\ 0 & 1 & 1 \\ 0 & 0 & 1 \end{pmatrix}.$$
We set: $B = A - I$.
(1) Find B^n for each n.
(2) Determine A^n.

Exercise 1.4.12. Is the sum of two non-nilpotent matrices non-nilpotent? What about their product?

Exercise 1.4.13. Let $A, B \in M_n$ be nilpotent, such that $AB = BA$. Show that $A + B$ and AB are both nilpotent.

Exercise 1.4.14. *A matrix B is called a cube root of a square matrix A if*
$$B^3 = A.$$
More generally, an nth root of A is a matrix B satisfying
$$B^n = A.$$
Show that the number of nonzero cube roots of a (complex) matrix is always a multiple of 3. Can this result be generalized to nth roots?

CHAPTER 2

Vector Spaces

2.1. Course Summary

2.1.1. Introduction to vector spaces.

DEFINITION 2.1.1. A vector space (or linear space) over $\mathbb{F}$ (or an $\mathbb{F}$-vector space) is a non-empty set E endowed with an addition, noted "+", and a scalar multiplication noted "·", such that:

- $\forall x, y \in E : x + y \in E$,
- $\forall x \in E, \forall \lambda \in \mathbb{F} : \lambda \cdot x \in E$,

and

$\mathbf{A_1}$: $x + (y + z) = (x + y) + z$, $\forall x, y, z \in E$.

$\mathbf{A_2}$: There is a vector in E denoted 0 such that:

$$\forall x \in E : x + 0 = 0 + x = x.$$

$\mathbf{A_3}$: For all x in E, there is a vector denoted $-x$ such that:

$$x + (-x) = (-x) + x = 0.$$

$\mathbf{A_4}$: $\forall x, y \in E : x + y = y + x$.

$\mathbf{M_1}$: $k \cdot (x + y) = k \cdot x + k \cdot y$, $\forall k \in K$, $\forall x, y \in E$.

$\mathbf{M_2}$: $(k + l) \cdot x = k \cdot x + l \cdot x$, $\forall k, l \in K$, $\forall x \in E$.

$\mathbf{M_3}$: $(kl) \cdot x = k \cdot (l \cdot x)$, $\forall k, l \in K$, $\forall x \in E$.

$\mathbf{M_4}$: $1 \cdot x = x$.

Remark. As it is customary, we shall write λx instead of $\lambda \cdot x$.

Here are some fundamental examples of vector spaces:

EXAMPLES 2.1.1.

(1) The set of all ordered n-tuples of real numbers, $\mathbb{R}^n$, is a vector space over $\mathbb{R}$. The operations of vector addition and scalar multiplication are defined as:

$$(x_1, \ldots, x_n) + (y_1, \ldots, y_n) = (x_1 + y_1, \ldots, x_n + y_n)$$

$$c(x_1, \ldots, x_n) = (cx_1, \ldots, cx_n),$$

where c is a real scalar. In particular, $\mathbb{R}$, with the usual operations of addition and multiplication, forms a vector space over $\mathbb{R}$.

(2) Similarly, we may define the vector space $\mathbb{C}^n$, which is the set of all ordered n-tuples of complex numbers:

$$\mathbb{C}^n = \{(z_1, z_2, \ldots, z_n) \mid z_i \in \mathbb{C}, \ i = 1, 2, \ldots, n\}.$$

The addition of two vectors $u = (u_1, u_2, \ldots, u_n)$ and $v = (v_1, v_2, \ldots, v_n)$ in $\mathbb{C}^n$ is defined as:

$$u + v = (u_1 + v_1, u_2 + v_2, \ldots, u_n + v_n),$$

while the scalar multiplication is defined, for a scalar $c \in \mathbb{C}$ and a vector $u = (u_1, u_2, \ldots, u_n)$ by:

$$cu = (cu_1, cu_2, \ldots, cu_n).$$

More generally, if $\mathbb{F}$ is a field, then $\mathbb{F}^n$ forms a vector space over $\mathbb{F}$ under component-wise addition and scalar multiplication, defined analogously to the operations in, e.g., $\mathbb{C}^n$. For example, $\mathbb{Q}^n$ is a $\mathbb{Q}$-vector space. Also, when p is a prime number, it is well known that the quotient ring $\mathbb{Z}/p\mathbb{Z}$, which we denote as $\mathbb{Z}_p$ in this book, is a field. Consequently, $\mathbb{Z}_p^n$ is a vector space over $\mathbb{Z}_p$.

Remark. It is noteworthy that $\mathbb{C}^n$ can also be regarded as a vector space over $\mathbb{R}$, where c above is restricted to being a real scalar.

Remark. In certain situations in $\mathbb{R}^n$ or $\mathbb{C}^n$, such as when $n = 2$, we commonly use both representations (a, b) and $\begin{pmatrix} a \\ b \end{pmatrix}$ to denote the same object.

(3) The set of all $n \times m$ matrices with real entries, denoted $M_{n \times m}(\mathbb{R})$, forms a vector space. The operations are matrix addition and scalar multiplication.

(4) The space of polynomials of degree at most n, with coefficients in a field $\mathbb{F}$, forms a vector space. Each such polynomial $p(X)$ can be written as

$$p(X) = a_0 + a_1 X + a_2 X^2 + \cdots + a_n X^n,$$

where $a_0, a_1, \ldots, a_n \in \mathbb{F}$. The operations of addition and scalar multiplication are defined in the usual way for polynomials. The vector space of polynomials of degree at most n is commonly denoted by $\mathbb{F}_n[X]$.

The notation $\mathbb{F}[X]$ represents the space of all polynomials with coefficients in the field $\mathbb{F}$, without any restriction on degree.

(5) (Cf. Exercise 2.3.2) Let E be a non-empty set and F be an $\mathbb{F}$-vector space, where $\mathbb{F}$ is a field. The set of functions from E to F is a vector space over $\mathbb{F}$ with respect to the operations:

$$(f + g)(x) = f(x) + g(x), \quad (\lambda f)(x) = \lambda f(x)$$

(for all $x \in E$).

(6) Let E and F be two vector spaces over the same field $\mathbb{F}$. Then $E \times F$ is a vector space over $\mathbb{F}$ with respect to the operations:

$$(x, y) + (x', y') = (x + x', y + y') \text{ and } \alpha(x, y) = (\alpha x, \alpha y).$$

In particular, $0_{E \times F} = (0_E, 0_F)$ and $-(x, y) = (-x, -y)$.

Remark. Similarly, we can endow the direct product $E_1 \times E_2 \times \cdots \times E_n$ with the structure of a vector space, where $E_1, E_2, \ldots, E_n$ are n vector spaces over $\mathbb{F}$.

(7) The set $\{0\}$ is a vector space, called the trivial vector space or zero vector space.

2.1.2. Subspaces.

We have the following fundamental and practical concept:

DEFINITION 2.1.2. Let E be a vector space over a field $\mathbb{F}$. Say that a nonempty subset F of E is a linear or vector subspace (or simply subspace) of E if F is a vector space under the same operations that already make E into a vector space.

We have the following equivalent definition:

DEFINITION 2.1.3. Let E be a vector space over a field $\mathbb{F}$ and let $F \subset E$. We say that F is a (linear) subspace of E if $F \neq \varnothing$ (in particular, F must at least contain 0_E) and

$$\forall \alpha, \beta \in \mathbb{F} : \forall x, y \in F : \ \alpha x + \beta y \in F.$$

Remark. A vector space always contains the zero vector, so we must always start by checking if $0_E \in F$. If this statement is true, we can't conclude anything and we need to further check the main property in the above definition. However, if that is false, meaning $0_E \notin F$, then F is not a (linear) subspace of E.

Remark. In practice, to show that a set is a vector space, we will always try to show that it is a linear subspace of a larger space that is already known to be a vector space.

EXAMPLES 2.1.2.

(1) $\mathbb{R}^*$ is not a subspace of $\mathbb{R}$, since $0 \notin \mathbb{R}^*$.

(2) The set $F = [0, 1]$ is not a linear subspace of $\mathbb{R}$, because, for example, $1, \frac{1}{2} \in F$, but $1 + \frac{1}{2} = \frac{3}{2} \notin F$.

(3) $\mathbb{Q}$ is not a subspace of $\mathbb{R}$, since $\sqrt{2} \cdot 1 = \sqrt{2} \notin \mathbb{Q}$.

(4) If E is a non-trivial vector space, then it has at least two subspaces: $\{0_E\}$ and E itself.

The following concepts are fundamental when working with vector spaces.

DEFINITION 2.1.4. A linear combination of a set of vectors v_1, $v_2, \ldots, v_k$ in a vector space E is an expression of the form:

$$a_1 v_1 + a_2 v_2 + \cdots + a_k v_k$$

where $a_1, a_2, \ldots, a_k$ are scalars from the field $\mathbb{F}$, and $v_1, v_2, \ldots, v_k$ are vectors in E.

DEFINITION 2.1.5. The span of a set of vectors $\{v_1, v_2, \ldots, v_k\}$ in a vector space E is the set of all possible linear combinations of these vectors. Formally, the span of $S := \{v_1, v_2, \ldots, v_k\}$, denoted by $\text{span}(S)$, is defined as:

$$\text{span}(S) = \{a_1 v_1 + a_2 v_2 + \cdots + a_k v_k : a_1, a_2, \ldots, a_k \in \mathbb{F}\}.$$

PROPOSITION 2.1.1. *(See Exercise 2.3.6) The span of a set of vectors is a vector space.*

2.1.3. Linear independence.

DEFINITION 2.1.6. A set of vectors $\{v_1, v_2, \ldots, v_k\}$ in a vector space E is said to be linearly independent (or free) if the only solution to the equation

$$a_1 v_1 + a_2 v_2 + \cdots + a_k v_k = 0$$

is

$$a_1 = a_2 = \cdots = a_k = 0.$$

If there exists any non-trivial solution (where at least one $a_i \neq 0$), the set is said to be linearly dependent.

Remark. A set of vectors $S = \{v_1, v_2, \ldots, v_k\}$ in a vector space is linearly *dependent* if there exist scalars $a_1, a_2, \ldots, a_k$, not all zero, such that

$$a_1 v_1 + a_2 v_2 + \cdots + a_k v_k = 0.$$

In other words, at least one vector in S can be expressed as a linear combination of the others.

EXAMPLES 2.1.3.

(1) In $\mathbb{R}^3$, the set $\{(1,0,0),(0,1,0),(0,0,1)\}$ is linearly independent.

(2) In $\mathbb{R}^3$, the set $\{(1,0,0),(0,1,0),(1,1,0)\}$ is linearly dependent because:

$$(1,1,0) = (1,0,0) + (0,1,0).$$

(3) Let E be a vector space and $a \in E$. The set $\{a\}$ is linearly independent if and only if $a \neq 0$.

DEFINITION 2.1.7. The rank of a family (or set) of vectors refers to the maximum number of linearly independent vectors within that family. In other words, it is the dimension of the subspace spanned by the vectors in the set.

The rank of a matrix is the maximum number of linearly independent rows or columns in the matrix.

2.1.4. Basis and dimension.

DEFINITION 2.1.8. A basis of a vector space E is a set of vectors $\{v_1, v_2, \ldots, v_k\}$ in E that satisfies:

- *Linear Independence:* The vectors are linearly independent.
- *Spanning:* The set spans E, i.e., *every* vector in E can be written as a linear combination of these vectors.

The dimension of a vector space E, denoted by $\dim E$, is the number of vectors in a basis of E. If this number is finite, we say that E is finite-dimensional.

If a vector space is not finite-dimensional, we say that it is infinite-dimensional.

The following theoretical result is useful.

THEOREM 2.1.2. *Let E be a vector space of dimension n and let B be a basis of E. Then*

(1) *Any other basis of E has n elements.*

(2) *A set S of n vectors in E is a basis of E if and only if S is either linearly independent or spans E.*

Remark. If $E = \{0\}$, the zero vector space, it has no basis. In this case, we define $\dim E = 0$.

When dealing with common vector spaces, such as $\mathbb{R}^n$, there are some simple and easily identifiable bases, called standard bases. Below are some common examples:

EXAMPLES 2.1.4.

(1) The standard basis of $\mathbb{R}^n$ consists of vectors where each vector has exactly one component equal to 1, and all other components are 0. More precisely, the standard basis of $\mathbb{R}^n$ is given by:

$$e_1 = (1, 0, \ldots, 0), e_2 = (0, 1, \ldots, 0), \ldots, e_n = (0, 0, \ldots, 1).$$

In particular, the standard basis for $\mathbb{R}^2$ is $\{(1,0),(0,1)\}$, whereas the standard basis for $\mathbb{R}^3$ is $\{(1,0,0),(0,1,0),(0,0,1)\}$. We thus see that $\dim \mathbb{R}^n = n$.

(2) The vector space $\mathbb{C}^n$ *over* $\mathbb{C}$ also has the following standard basis:

$$e_1 = (1, 0, \ldots, 0), e_2 = (0, 1, \ldots, 0), \ldots, e_n = (0, 0, \ldots, 1).$$

So, $\dim \mathbb{C}^n = n$ (when $\mathbb{C}^n$ is a $\mathbb{C}$-vector space).

(3) The standard basis of $\mathbb{R}_n[X]$ is $\{1, X, X^2, \ldots, X^n\}$. Thus,

$$\dim \mathbb{R}_n[X] = n + 1.$$

(4) The standard basis of $M_2(\mathbb{R})$ is given by

$$\left\{ \begin{pmatrix} 1 & 0 \\ 0 & 0 \end{pmatrix}, \begin{pmatrix} 0 & 1 \\ 0 & 0 \end{pmatrix}, \begin{pmatrix} 0 & 0 \\ 1 & 0 \end{pmatrix}, \begin{pmatrix} 0 & 0 \\ 0 & 1 \end{pmatrix} \right\},$$

which yields $\dim M_2(\mathbb{R}) = 4$. More generally, $\dim M_n(\mathbb{R}) = n^2$, and, even more generally, $\dim M_{n \times m}(\mathbb{R}) = n \times m$. For instance, the canonical basis of $M_{2 \times 3}(\mathbb{R})$ (of dimension 6) is constituted of:

$$\begin{pmatrix} 1 & 0 & 0 \\ 0 & 0 & 0 \end{pmatrix}, \begin{pmatrix} 0 & 1 & 0 \\ 0 & 0 & 0 \end{pmatrix}, \begin{pmatrix} 0 & 0 & 1 \\ 0 & 0 & 0 \end{pmatrix},$$

$$\begin{pmatrix} 0 & 0 & 0 \\ 1 & 0 & 0 \end{pmatrix}, \begin{pmatrix} 0 & 0 & 0 \\ 0 & 1 & 0 \end{pmatrix} \text{ and } \begin{pmatrix} 0 & 0 & 0 \\ 0 & 0 & 1 \end{pmatrix}.$$

THEOREM 2.1.3. *Let E be a finite-dimensional vector space, and let $F \subseteq E$ be a subspace of E. Then, F is finite-dimensional, and $\dim F \leq \dim E$. Furthermore, if $\dim E = \dim F$, then $E = F$.*

In the end, we have yet another fundamental result: the Incomplete Basis Theorem.

THEOREM 2.1.4. *Let E be a vector space of dimension n. For every linearly independent set $\{u_1, u_2, \ldots, u_p\}$ (with $p < n$), we can find $q = n - p$ vectors in E, denoted by $\{v_1, v_2, \ldots, v_q\}$, such that $\{u_1, u_2, \ldots, u_p, v_1, v_2, \ldots, v_q\}$ is a basis of E.*

2.1.5. Complementary subspaces.

DEFINITION 2.1.9. Let E be a vector space, and let F and G be subspaces of E. We say that F and G are complementary vector spaces (or supplementary vector spaces) if every vector $v \in E$ can be uniquely written as:

$$v = f + g, \quad \text{with } f \in F \text{ and } g \in G.$$

When this holds, we write:

$$E = F \oplus G,$$

and E is called the direct sum of F and G. In this case, we say that F is a complement (or supplement) of G.

Equivalent definitions. The subspaces F and G are complementary in E if and only if any of the following equivalent conditions hold:

(1) $F \cap G = \{0\}$ and $F + G = E$, where:

$$F + G = \{f + g \mid f \in F, g \in G\}.$$

(2) $F \cap G = \{0\}$, and $\dim F + \dim G = \dim E$ (if E is finite-dimensional).

EXAMPLE 2.1.1. Let $E = \mathbb{R}^3$, $F = \operatorname{span}\{(1,0,0),(0,1,0)\}$, and $G = \operatorname{span}\{(0,0,1)\}$. Then:

$$\mathbb{R}^3 = F \oplus G.$$

2.2. True or False

Questions. Answer the following questions, providing justifications for your answers:

(1) A (linear) subspace is a vector space.

(2) $\mathbb{R}^2$ is a subspace of $\mathbb{R}^3$.

(3) The intersection of two subspaces (of the same vector space E) is never empty.

(4) The intersection of two subspaces of the same vector space E is always a vector space.

(5) The union of two subspaces is always a vector space.

(6) Let E and F be two vector spaces over the same field $\mathbb{F}$. Then, $E + F$ is a vector space over $\mathbb{F}$, where

$$E + F = \{x + y : x \in E, y \in F\}.$$

(7) $\mathbb{R}$ is vector space over $\mathbb{C}$, and $\mathbb{C}$ is a vector space over $\mathbb{R}$.

(8) Let F, G, and H be (linear) subspaces of a vector space E. If $F + H = G + H$, then $F = G$.

(9) Let E be a vector space over a field $\mathbb{F}$. Then $E - E = \{0_E\}$.

(10) Let E be a vector space over a field $\mathbb{F}$. Then $aE = E$ for any $a \in \mathbb{F}$, where $aE = \{ax : x \in E\}$.

(11) Let $\{e_1, \ldots, e_n\}$ be a family of pairwise non-collinear vectors. Then, $\{e_1, \ldots, e_n\}$ is free.

(12) Let $e_1, \ldots, e_n$ and $f_1, \ldots, f_n$ be two linearly independent families of vectors. Then, $\{e_1 + f_1, \ldots, e_n + f_n\}$ is free.

(13) Let $\{e_1, e_2, \ldots, e_n\}$ be a linearly independent family of vectors. Then, $\{-e_1, e_2, \ldots, e_n\}$ too is a linearly independent family.

(14) Let E be a vector space of dimension n. Then E can have infinitely many bases.

(15) The dimension of $\mathbb{R}$ over itself is one.

(16) Let $\mathbb{C}$ be the usual vector space over $\mathbb{R}$. What is its dimension? Will its dimension remain the same when considered as a vector space over $\mathbb{C}$?

(17) The dimension of $\mathbb{R}$, when it is considered as a vector space over $\mathbb{Q}$, is infinite.

(18) Let E and F be two subspaces of $\mathbb{R}^3$ such that $\dim E = \dim F = 2$. Then $E \cap F \neq \{0\}$.

(19) Assume that F and G are two different subspaces of the same vector space E. Suppose that $\dim E = 6$ and $\dim F = \dim G = 4$. What are the possible dimensions of $F \cap G$?

(20) Every vector space always has a basis.

(21) Complementary subspaces are not always unique.

(22) Let A and B be two square matrices of order n. Then

$$\text{rank}(AB) = \text{rank}(BA).$$

Answers.

(1) **True.** A subspace is indeed a vector space. A subspace inherits the operations of vector addition and scalar multiplication from its parent vector space and satisfies all the axioms that define a vector space. It is worth noting that every vector space can be considered a subspace of itself.

(2) **False!** $\mathbb{R}^2$ is not a subspace of $\mathbb{R}^3$ because the vectors in $\mathbb{R}^2$ have two components, while the vectors in $\mathbb{R}^3$ have three components. A subspace of a vector space must consist of vectors with the same dimension as the vectors in the vector space itself. In other words, for $\mathbb{R}^2$ to be a subspace of $\mathbb{R}^3$, its vectors would need to be of the form (x, y, z), where each vector has three components. However, the vectors in $\mathbb{R}^2$ are of the form (x, y) with only two components. Once that's clear, readers can further observe that $\mathbb{R}^2$ cannot even be considered

a subset of $\mathbb{R}^3$. However, if we "identify" (whatever this word could mean to you) $\mathbb{R}^2$ with $\mathbb{R}^2 \times \{0\}$, where $\mathbb{R}^2 \times \{0\} = \{(x, y, 0) : x, y \in \mathbb{R}\}$, then this way we can say that this "copy" of $\mathbb{R}^2$ is a subspace of $\mathbb{R}^3$.

(3) **True.** The reason is that every subspace of a vector space must contain the zero vector of the ambient vector space, ensuring that their intersection is always nonempty.

(4) **True.** Let us prove the result. Let E be a vector space, and let F and G be two subspaces of E. We need to show that $F \cap G$ is also a subspace of E.

First, since both F and G are subspaces, they contain the zero vector of E. Hence, $0 \in F \cap G$.

Next, let $x, y \in F \cap G$ and $a, b \in \mathbb{F}$ (where $\mathbb{F}$ is the underlying field of E). So, $x, y \in F$ and $x, y \in G$. Since F and G are subspaces, $ax + by \in F$ and $ax + by \in G$. It follows that $ax + by \in F \cap G$. Therefore, $F \cap G$ is closed under addition and scalar multiplication, proving that it is a subspace of E.

Remark. This result extends to arbitrary (finite or infinite) intersections of subspaces. (See Exercise 2.4.3.)

(5) **False!** For instance, consider the sets:
$$E = \{(x, x) : x \in \mathbb{R}\} \quad \text{and} \quad F = \{(y, -y) : y \in \mathbb{R}\}.$$
Clearly, E is spanned by the vector $(1, 1)$, so it is a subspace of $\mathbb{R}^2$. Similarly, F is a subspace of $\mathbb{R}^2$ because it is spanned by the vector $(1, -1)$. However, the union $E \cup F$ is not a subspace of $\mathbb{R}^2$. Indeed, $(1, 1) \in E \cup F$ and $(1, -1) \in E \cup F$, yet
$$(1, 1) + (1, -1) = (2, 0) \notin E \cup F.$$

THEOREM 2.2.1. *If E is a vector space and F and G are two subspaces of E, then*
$$F \cup G \text{ is a subspace of } E \iff F \subset G \text{ or } G \subset F.$$

As a consequence, we have the following corollary:

COROLLARY 2.2.2. *If E is a vector space and F and G are two subspaces of E, then:*
$$F \neq E \text{ and } G \neq E \implies F \cup G \neq E.$$

(6) **False!** For example, $\mathbb{R}^2$ and $\mathbb{R}^3$ are two vector spaces over the field $\mathbb{R}$, but $\mathbb{R}^2 + \mathbb{R}^3$ cannot be a vector space. There is a main obstacle preventing $\mathbb{R}^2 + \mathbb{R}^3$ from being a vector space, which is the fact that the elements of $\mathbb{R}^2$ and $\mathbb{R}^3$ have different

dimensions; therefore, we cannot add vectors from $\mathbb{R}^2$ and $\mathbb{R}^3$ in the usual sense. Furthermore, it is unclear how to define a zero vector for $\mathbb{R}^2 + \mathbb{R}^3$! This counterexample shows that the sum of two vector spaces may not form a vector space when they do not share a common ambient space.

The concept of "sum" is more meaningful for subspaces within a common larger vector space rather than for arbitrary vector spaces. More precisely, when E and F are subspaces of the same vector space, say V, then $E + F$ is a subspace of V. Hence, $E + F$ is itself a vector space. The proof is left to the interested reader. In this context, we have the following important result, often referred to as the dimension formula:

$$\dim(E + F) = \dim E + \dim F - \dim(E \cap F).$$

In particular,

$$\dim(E \oplus F) = \dim E + \dim F.$$

(7) The set of complex numbers $\mathbb{C}$ forms a vector space over $\mathbb{R}$ because it satisfies all the necessary axioms for a vector space when considering $\mathbb{R}$ as the field of scalars, and details are left to readers.

However, $\mathbb{R}$ cannot be considered as a vector space over $\mathbb{C}$. Indeed, if $\mathbb{R}$ were a vector space over $\mathbb{C}$, then multiplying a real number by a complex number would have to yield a real number, which is untrue. For example, if we take the real number 1 (from $\mathbb{R}$) and multiply it by the complex number i (from $\mathbb{C}$), we would get $1 \cdot i = i$, which is not an element of $\mathbb{R}$.

(8) **False!** Let F and G be two subspaces of E such that $F \neq G$ and set $H = E$. Then

$$F + H = F + E = E = G + E = G + H,$$

yet we have chosen $F \neq G$.

Remark. Let us quickly demonstrate that $F + E = E$.

Suppose $x + y \in F + E$, where $x \in F$ and $y \in E$. Since every element of F is also in E, we have $x \in E$. Since E is a vector space and is closed under addition, it follows that $x + y \in E$. Thus, $F + E \subset E$.

Conversely, if $x \in E$, we can write $x = 0_E + x$, where 0_E is the zero vector in E. Since $0_E \in F$, this shows that $x \in F + E$. Therefore, we conclude that $F + E = E$.

(9) **False!** (Unless $E = \{0_E\}$.) We start by providing a counterexample. If $E = \mathbb{R}$, then $\mathbb{R} - \mathbb{R} = \{x - y : x, y \in \mathbb{R}\} \neq \{0\}$ as, e.g., $-1 = 1 - 2 \in \mathbb{R} - \mathbb{R}$. In fact, let us show that $E - E = E$. Let $x, y \in E$. Then

$$E - E \ni x - y \in E,$$

where $E - E \ni x - y$ by definition and $x - y \in E$ because E is a vector space. So, we have shown that $E - E \subset E$. Furthermore, if $x \in E$, then we can express x as $x - 0_E$, implying that $x \in E - E$, and thus, $E \subset E - E$. Consequently, we have proven that $E - E = E$.

(10) **False!** (Unless $E = \{0_E\}$.) As a counterexample, observe that $0\mathbb{R} = \{0\} \neq \mathbb{R}$. However, when $a \neq 0$, then $aE = E$. In order to demonstrate this, let us start with $x \in E$. Since E is a vector space, we have $a^{-1}x \in E$. Therefore, $x = a(a^{-1}x) \in aE$, which implies $E \subset aE$. Conversely, if $y \in aE$, then $y = ax$, where $x \in E$. As E is a vector space, it follows that $ax \in E$.

We leave it to the reader to show that if E is a vector space, then $aE + bE = E$, provided $a, b \in \mathbb{F}$ with at least one of a or b being nonzero.

(11) **False!** Consider in $\mathbb{R}^2$ the three vectors $e_1 = (1, 0)$, $e_2 = (0, 1)$, and $e_3 = (1, 1)$. These vectors are pairwise non-collinear. However, the set $\{e_1, e_2, e_3\}$ is not free as its cardinal is three while in a vector space of dimension two.

(12) **False!** For a counterexample, let $\{e_1, e_2, \ldots, e_n\}$ be a free set, then consider $\{-e_1, -e_2, \ldots, -e_n\}$, which stays a free set. However,

$$\{e_1 - e_1, e_2 - e_2, \ldots, e_n - e_n\} = \{0, 0, \ldots, 0\}$$

is obviously not a free set.

(13) **True.** Let $\{e_1, e_2, \ldots, e_n\}$ be a free set. To demonstrate that $\{-e_1, e_2, \ldots, e_n\}$ remains free, take scalars $\alpha_1, \alpha_2, \ldots, \alpha_n$ such that $\alpha_1(-e_1) + \alpha_2 e_2 + \cdots + \alpha_n e_n = 0$, or equivalently, $(-\alpha_1)e_1 + \alpha_2 e_2 + \cdots + \alpha_n e_n = 0$. Since $e_1, e_2, \ldots, e_n$ are linearly independent, it ensues that $-\alpha_1 = \alpha_2 = \cdots = \alpha_n = 0$, showing that the vectors $-e_1, e_2, \ldots, e_n$ are linearly independent, as wished.

(14) **True.** For example, $\mathbb{R}^2$ has infinitely many distinct bases given by $\{(a, 0), (0, b)\}_{a,b \in \mathbb{R}^*}$, as it might be checked.

(15) **True.** To prove this result, let x be any real number. It is plain that $x = x \cdot 1$, indicating that the set $\{1\}$ spans $\mathbb{R}$. If a is

such that $a \cdot 1 = 0$, then obviously $a = 0$, making the set $\{1\}$ free. Thus, $\{1\}$ is a basis of $\mathbb{R}$, and so $\dim \mathbb{R} = \operatorname{card}\{1\} = 1$.

(16) If $\mathbb{C}$ is considered a vector space over $\mathbb{R}$, then its dimension, noted $\dim_{\mathbb{R}} \mathbb{C}$ is two. To see why, take any $z \in \mathbb{C}$. Then $z = x + iy = x \cdot 1 + y \cdot i$, which shows that the family $\{1, i\}$ spans $\mathbb{C}$. To show that this family is free, let $\alpha, \beta \in \mathbb{R}$ be such that $\alpha \cdot 1 + \beta \cdot i = 0$. Through basic properties of complex numbers, we immediately deduce that $\alpha = \beta = 0$. Thus, $\{1, i\}$ is a basis of $\mathbb{C}$ over $\mathbb{R}$, and thus $\dim_{\mathbb{R}} \mathbb{C} = 2$.

If $\mathbb{C}$ is considered a vector space over $\mathbb{C}$, then $\dim \mathbb{C} = 1$. The proof resembles that of showing the dimension of $\mathbb{R}$ over $\mathbb{R}$ is one, and is therefore omitted.

(17) **True.** The dimension of the vector space $\mathbb{R}$ over the field $\mathbb{Q}$ is infinite. For a proof, see Exercise 2.4.18.

(18) **True.** Suppose that $E \cap F = \{0\}$. Then:

$$\dim(E + F) = \dim E + \dim F - \underbrace{\dim(E \cap F)}_{0}$$

$$= \dim E + \dim F = 4.$$

But $E + F$ is a subspace of $\mathbb{R}^3$, so its dimension is less than or equal to 3, so its dimension is not equal to 4! So we have: $E \cap F \neq \{0\}$.

(19) Since F and G are different, we have:

$$F \subsetneq F + G \text{ and } G \subsetneq F + G$$

(i.e., we have proper or strict inclusions). So, $\dim(F + G) > 4$. On the other hand,

$$\dim(F + G) \leq \dim E = 6.$$

Hence,

$$\dim(F + G) = 5 \text{ or } \dim(F + G) = 6.$$

But

$$\dim(F + G) = \dim F + \dim G - \dim(F \cap G),$$

that is:

$$\dim(F \cap G) = \dim F + \dim G - \dim(F + G)$$
$$= 4 + 4 - \dim(F + G),$$

and

$$\dim(F \cap G) = 8 - \dim(F + G).$$

Thus,

$$\dim(F \cap G) = 3 \text{ or } \dim(F \cap G) = 2.$$

(20) If a vector space E is trivial (i.e., $E = \{0\}$), it does not have a basis. However, if E is nontrivial and is either finite-dimensional or infinite-dimensional, then E has a basis.

(21) **True.** Take $E = \mathbb{R}^2$ and let $F = \{(x, 0) : x \in \mathbb{R}\}$. We show that F has, e.g., two complements, given by $G_1 = \{(0, y) : y \in \mathbb{R}\}$ and $G_2 = \{(x, x) : x \in \mathbb{R}\}$. It is clear that $\dim F = \dim G_1 = \dim G_2 = 1$. Thus, $\dim E = \dim F + \dim G_1$ and $\dim E = \dim F + \dim G_2$. Since $F \cap G_1 = \{0\}$, it ensues that $E = F \oplus G_1$. Since, $F \cap G_1 = \{0\}$, we too have $E = F \oplus G_1$. Therefore, F and G_1 are complementary, as are F and G_2.

(22) **False!** Let:

$$A = \begin{pmatrix} 0 & 0 \\ 1 & 0 \end{pmatrix} \text{ and } B = \begin{pmatrix} 1 & 0 \\ 0 & 0 \end{pmatrix}.$$

Then:

$$AB = \begin{pmatrix} 0 & 0 \\ 1 & 0 \end{pmatrix} \text{ and } BA = \begin{pmatrix} 0 & 0 \\ 0 & 0 \end{pmatrix}.$$

So trivially:

$$\text{rank}(AB) = 1 \neq \text{rank}(BA) = 0.$$

2.3. Exercises with Solutions

Exercise 2.3.1. Let E be a vector space. Simplify, when possible, the following expressions:

(1) $x + y - 2(x - y)$,
(2) $-x + xy + x^2$,
(3) $x2 + 3x - y + 2(y - x)$,
(4) $x/y + 9y$,

where $x, y \in E$.

Solution 2.3.1.

(1) We have:

$$x + y - 2(x - y) = x + y - 2x + 2y = -x + 3y.$$

(2) This is not a well-defined expression in a vector space because the product of two vectors, such as xy, or even x^2, is not defined in vector spaces.

(3) This expression is not accepted in a vector space because the scalar should be written before the vector, i.e., we should write $2x$ rather than $x2$. In the context of vector spaces, it is crucial to maintain the distinction between scalar multiplication and other operations to ensure clarity, consistency, and adherence to standard mathematical conventions in vector spaces. One exception occurs when the vector space coincides with the field (e.g., the vector space $\mathbb{R}$ over $\mathbb{R}$).

(4) The expression appears to involve the division of two vectors. However, division by a vector is not defined in the context of vector spaces.

Exercise 2.3.2. Show that the set E of functions from $\mathbb{R}$ to $\mathbb{R}$ is a vector space over $\mathbb{R}$ with respect to the operations:

$$(f+g)(x) = f(x) + g(x), \quad (\lambda f)(x) = \lambda f(x),$$

where $f, g \in E$ and $\lambda \in \mathbb{R}$.

Solution 2.3.2.

- Let $f, g, h \in E$. For all x, we have:

$$\begin{aligned}
[(f+g)+h](x) &= [(f+g)(x)] + h(x) \\
&= (f(x) + g(x)) + h(x) \\
&= f(x) + (g(x) + h(x)) \\
&= [f + (g+h)](x)
\end{aligned}$$

because the values of the functions lie in $\mathbb{R}$, and $+$ is associative over $\mathbb{R}$. So:

$$(f+g)+h = f+(g+h).$$

- The zero function, noted $\mathbf{0}$, satisfies:

$$\forall f \in E: \ f + \mathbf{0} = \mathbf{0} + f = f$$

because the scalar "0" is the identity element of $\mathbb{R}$ with respect to addition. Hence

$$\forall x \in \mathbb{R}: \ f(x) + 0 = 0 + f(x) = f(x).$$

- The inverse of $f \in E$ is $-f$ (which belongs to E) because:

$$\forall x \in \mathbb{R}: \ f(x) + (-f(x)) = (-f(x)) + f(x) = 0.$$

- Let $f, g \in E$. Since addition is commutative in $\mathbb{R}$, we get:

$$\forall x \in \mathbb{R}: \ f(x) + g(x) = g(x) + f(x),$$

i.e., $f + g = g + f$.

- Let $f, g \in E$ and $\lambda \in \mathbb{R}$. Since multiplication is distributive (on the left) with respect to addition in $\mathbb{R}$, we have:

$$\forall x \in \mathbb{R}: \ \lambda(f + g)(x) = \lambda(f(x) + g(x)) = \lambda f(x) + \lambda g(x),$$

  i.e., $\lambda(f + g) = \lambda f + \lambda g$.
- Let $f \in E$; $\lambda, \mu \in \mathbb{R}$. We have:

$$\forall x \in \mathbb{R}: \ (\lambda + \mu)f(x) = \lambda f(x) + \mu f(x),$$

  because multiplication is distributive (on the right) with respect to addition in $\mathbb{R}$. Hence, $(\lambda + \mu)f = \lambda f + \mu f$.
- Let $f \in E$; $\lambda, \mu \in \mathbb{R}$. Since multiplication is associative in $\mathbb{R}$, we have:

$$\forall x \in \mathbb{R}: \ (\lambda\mu)(f(x)) = \lambda(\mu f(x)),$$

  i.e., $(\lambda\mu)(f) = \lambda(\mu f)$.
- Let $\mathbf{1}$ be the constant function that is equal to 1 on $\mathbb{R}$. Let $f \in E$. Then:

$$\forall x \in \mathbb{R}: \ (\mathbf{1}f)(x) = 1 \cdot f(x) = f(x)$$

  (since 1 is the identity element for multiplication in $\mathbb{R}$). Thus:

$$\mathbf{1}f = f.$$

Exercise 2.3.3. Let $F = C(\mathbb{R}, \mathbb{R})$ be the set of *continuous* functions from $\mathbb{R}$ to $\mathbb{R}$. Prove that $(F, +, \cdot)$ is a vector space over the field $\mathbb{R}$ where the algebraic laws are defined by:

$$(f + g)(x) = f(x) + g(x), \ (\lambda f)(x) = \lambda f(x),$$

$f, g \in F, \lambda \in \mathbb{R}$.

Solution 2.3.3. Rather than proving that F is a vector space in its own right, we will demonstrate that it is a vector subspace of the already established vector space E, which appeared in Exercise 2.3.2.

First, we observe that F is non-empty, since the zero function (which maps every real number to 0) is continuous from $\mathbb{R}$ to $\mathbb{R}$, and hence belongs to F.

Now, let $f, g \in F$ and $\alpha, \beta \in \mathbb{R}$. Using basic properties of continuous functions, we know that the linear combination $\alpha f + \beta g$ is also continuous on $\mathbb{R}$. Therefore, $\alpha f + \beta g \in F$.

Thus, F is closed under addition and scalar multiplication, and since it contains the zero function, we conclude that F is a subspace of E. Consequently, F is itself a vector space.

Exercise 2.3.4. Let $E = \mathbb{R}^2$. Let $u = (x, y)$, $v = (x', y') \in E$ and $\lambda \in \mathbb{R}$. We define two algebraic laws "+" and "·" as follows:

$$u + v = (x + x', y + y') \text{ and } \lambda \cdot u = (\lambda x, 0).$$

Is $(E, +, \cdot)$ a vector space over $\mathbb{R}$?

Solution 2.3.4. No! For instance, let $u = (2, 1) \in \mathbb{R}^2$. The observation

$$1 \cdot u = 1 \cdot (2, 1) = (1 \times 2, 0) = (2, 0) \neq (2, 1) = u$$

is sufficient to conclude that $(E, +, \cdot)$ is not a vector space over $\mathbb{R}$.

Exercise 2.3.5. Let E be a vector space over a field $\mathbb{F}$. Prove for all $x \in E$ and $k \in \mathbb{F}$ that:
  (1) $k0 = 0$.
  (2) $0x = 0$.
  (3) $kx = 0 \Rightarrow k = 0 \ \lor \ x = 0$.
  (4) $(-k)x = k(-x) = -kx$.

Solution 2.3.5. Let us use Definition 2.1.1. Let $x \in E$ and $k \in \mathbb{F}$.

  (1) We have:

$$0 = 0 + 0 \text{ on } E, \text{ by } A_2$$
$$\Rightarrow k0 = k(0 + 0) = k0 + k0 \text{ by } M_1$$
$$\Rightarrow \underbrace{k0 + (-k0)}_{=0 \text{ by } A_3} = k0 + \underbrace{k0 + (-k0)}_{=0 \text{ by } A_3}$$
$$\Rightarrow 0 = \underbrace{k0 + 0}_{=0 \text{ by } A_2}.$$

Thus, $k0 = 0$, as desired.

  (2) We have: $0 + 0 = 0$ on $\mathbb{F}$. Then,

$$0x = (0 + 0)x = 0x + 0x \text{ by } M_2$$
$$\Rightarrow \underbrace{0x + (-0x)}_{=0 \text{ by } A_3} = 0x + \underbrace{0x + (-0x)}_{=0 \text{ by } A_3}$$
$$\Rightarrow 0 = 0x + 0 = 0x \text{ by } A_2.$$

  (3) Assume that $kx = 0$ with $k \neq 0$. Then, $k^{-1}k = 1$, and so

$$x = 1x \text{ by } M_4$$
$$= (k^{-1}k)x = k^{-1}(kx) \text{ by } M_3$$
$$= k^{-1}(0) = 0 \text{ (property on } \mathbb{F}).$$

(4) We know that: $k + (-k) = 0$ ($\mathbb{F}$ is a group). So,

$$0 = 0x = (k + (-k))x = kx + (-k)x \text{ by } M_2$$

$$\Rightarrow \underbrace{(-kx) + 0}_{=-kx \text{ by } A_2} = \underbrace{(-kx) + kx}_{=0 \text{ by } A_3} + (-k)x = \underbrace{0 + (-k)x}_{=(-k)x \text{ by } A_2}$$

Hence, $(-k)x = -kx$. We show, in a similar way and using $x + (-x) = 0$ (which comes from A_3), that $k(-x) = -kx$.

Exercise 2.3.6. Consider a set of vectors $S = \{v_1, v_2, \ldots, v_k\}$ in a vector space E. Show that span(S) is a vector space.

Solution 2.3.6. To show that span(S) is a vector space, it suffices to prove that it is a subspace of E. Recall that

$$\text{span}(S) = \{a_1 v_1 + a_2 v_2 + \cdots + a_k v_k : a_1, a_2, \ldots, a_k \in \mathbb{F}\}.$$

We verify the subspace criteria:

- Setting $a_1 = a_2 = \cdots = a_k = 0$, we obtain $0 \in \text{span}(S)$. Thus, span(S) contains the zero vector.
- Let $u, v \in \text{span}(S)$. Then $u = a_1 v_1 + a_2 v_2 + \cdots + a_k v_k$ and $v = b_1 v_1 + b_2 v_2 + \cdots + b_k v_k$ for some $a_i, b_i \in \mathbb{F}$. Their sum is:

$$u + v = (a_1 + b_1)v_1 + (a_2 + b_2)v_2 + \cdots + (a_k + b_k)v_k.$$

  Since $a_i + b_i \in \mathbb{F}$, it follows that $u + v \in \text{span}(S)$.
- Let $u \in \text{span}(S)$ and $\alpha \in \mathbb{F}$. Then $u = a_1 v_1 + a_2 v_2 + \cdots + a_k v_k$ for some $a_i \in \mathbb{F}$. Scalar multiplication gives:

$$\alpha u = (\alpha a_1)v_1 + (\alpha a_2)v_2 + \cdots + (\alpha a_k)v_k.$$

  Since $\alpha a_i \in \mathbb{F}$, it follows that $\alpha u \in \text{span}(S)$.

Thus, span(S) is indeed a subspace of E, and therefore it is a vector space.

Exercise 2.3.7. Indicate which of the following sets constitute a subspace of the real vector space $\mathbb{R}^2$:

(1) $A = \{(x, y) : x \geq 0\}$ and $B = \{(x, y) : x \leq 0\}$, and $A \cup B$.
(2) $C = \{(x, y) : x = 3y + 1\}$.
(3) The unit circle of center $(0, 0)$, denoted by D.
(4) $E = \mathbb{Z} \times \mathbb{R}$.
(5) $F = \{(0, 0)\}$.
(6) $G = \{(x(1, 2) : x \in \mathbb{R}\}$.
(7) $H = \{(x(1, 2) : x \in \mathbb{Q}\}$.

Solution 2.3.7.

(1) A is not a subspace because $(1,2) \in A$ since $1 \geq 0$, yet $-1(1,2) = (-1,2) \notin A$ since $-1 \not\geq 0$. Now, B is not a subspace because $(-2,0) \in B$ since $-2 \leq 0$, yet $-3(-2,0) = (6,0) \notin B$ since $6 \not\leq 0$.

Lastly, since $A \cup B = \mathbb{R}^2$, $A \cup B$ *is* a vector space.

Remark. We observe that the union of two sets that are not subspaces can still form a vector space.

(2) Since $(0,0) \notin C$ because $0 \neq 3 \times 0 + 1$, we conclude that C is not a vector space.

(3) Recall that $D = \{(x,y) \in \mathbb{R}^2 : x^2 + y^2 = 1\}$. One reason why D is not a subspace is that $(0,0) \notin D$ because $0^2 + 0^2 = 0 \neq 1$.

(4) Recall that the elements of E are of the form (n,x) with $n \in \mathbb{Z}$ and $x \in \mathbb{R}$. E is not a vector space because $(1, \sqrt{3}) \in E$ since $1 \in \mathbb{Z}$, yet $1/2(1, \sqrt{3}) = (1/2, \sqrt{3}/2) \notin E$ (as $\frac{1}{2} \notin \mathbb{Z}$).

(5) F is a subspace of $\mathbb{R}^2$ because it consists of only the zero vector $(0,0)$, which satisfies all the conditions for being a subspace.

(6) G is a subspace because it is spanned by the single vector $(1,2)$.

(7) H is not a subspace because it is not closed under scalar multiplication. Specifically, we cannot say that H is spanned by a single vector because the field of scalars is $\mathbb{R}$, not $\mathbb{Q}$. Note that H is a subspace of $\mathbb{Q}^2$, and in this case, it is spanned by a vector over the rational numbers.

That being said, consider the vector $(2,4) \in H$, corresponding to $x = 2 \in \mathbb{Q}$. However, if we multiply $(2,4)$ by $\sqrt{2} \in \mathbb{R}$, we get the vector $(2\sqrt{2}, 4\sqrt{2})$, which is not in H. This is because there is no $x \in \mathbb{Q}$ such that $(x, 2x) = (2\sqrt{2}, 4\sqrt{2})$.

Thus, H is not closed under scalar multiplication, and therefore, it cannot be a subspace.

Exercise 2.3.8. Is $F = \{(x,y) \in \mathbb{Z}_5^2 : x^2 + y^2 = 0\}$ a subspace of the vector space $\mathbb{Z}_5^2$?

Solution 2.3.8. First, recall that calculations are performed modulo 5. While it is clear that $(0,0) \in \mathbb{Z}_5^2$, F is not a subspace of $\mathbb{Z}_5^2$, as it is not closed under addition. For example, consider $(1,2), (2,1) \in F$, since $1^2 + 2^2 = 2^2 + 1^2 = 5 \equiv 0 \mod 5$. However, $(1,2) + (2,1) = (3,3)$, and we have $3^2 + 3^2 = 18 \equiv 3 \mod 5$, which is not equal to 0. Therefore, $(3,3) \notin F$, showing that F is not closed under addition.

Exercise 2.3.9. Which of the following sets are vector subspaces, and which are not?

(1) $E = \{(x, y, z) \in \mathbb{R}^3 : x + y - z = 1\}$,

(2) $E = \{(x, y, z) \in \mathbb{R}^3 : x + y^2 - z = 0\}$,

(3) $E = \{(x, y, z) \in \mathbb{R}^3 : x + y - z = 0\}$,

(4) $E = \{(x, y, z) \in \mathbb{R}^3 : x + \sqrt{|y|} - z = 0\}$,

(5) $E = \{(x, y, z) \in \mathbb{R}^3 : x + 2y - |z| = 0\}$,

(6) $E = \{(x, y, z) \in \mathbb{R}^3 : x^2 + y^2 + z^2 = 0\}$,

(7) $E = \{(x, y) \in \mathbb{R}^2 : xy = 0\}$,

(8) $E = \{(x, y, z, t) \in \mathbb{R}^4 : x + y = 0, z + 2t = 0\}$,

(9) $E = \{(x, y) \in \mathbb{R}^2 : x = 0\}$,

(10) $E = \{(x, y, z) \in \mathbb{R}^3 : x = y = z\}$,

(11) $E = \{(x, y, z) \in \mathbb{R}^3 : x = z\}$,

(12) $E = \{(x, y, z, t) \in \mathbb{R}^4 : y = z\}$.

Solution 2.3.9.

(1) The set is not a vector space because $(0, 0, 0) \notin E$, as

$$0 + 0 - 0 = 0 \neq 1.$$

(2) In this case, $0_{\mathbb{R}^3} \in E$, yet E is not a vector space. We provide a counterexample. It is clear that $(0, 2, 4) \in E$ since $0 + 2^2 - 4 = 0$, and $(0, 1, 1) \in E$ since $0 + 1^2 - 1 = 0$. However,

$$(0, 2, 4) + (0, 1, 1) = (0, 3, 5) \notin E,$$

because $0 + 3^2 - 5 = 4 \neq 0$. Therefore, E is not a vector space.

(3) In this case, E is a vector space. First, we can see that $(0, 0, 0) \in E$, as $0 + 0 - 0 = 0$. Next, let $(x, y, z), (x', y', z') \in E$, meaning $x + y - z = 0$ and $x' + y' - z' = 0$. Let $\alpha, \beta \in \mathbb{R}$. We check that $\alpha(x, y, z) + \beta(x', y', z') \in E$, i.e., we verify that

$$(\alpha x + \beta x', \alpha y + \beta y', \alpha z + \beta z') \in E.$$

We have:

$$(\alpha x + \beta x') + (\alpha y + \beta y') - (\alpha z + \beta z')$$
$$= \alpha x + \beta x' + \alpha y + \beta y' - \alpha z - \beta z'$$
$$= \alpha(x + y - z) + \beta(x' + y' - z')$$
$$= \alpha \cdot 0 + \beta \cdot 0 = 0.$$

Thus, E is a subspace.

(4) The set E is not a subspace. The interested reader can then consider the elements $(0, 4, 2)$ and $(-1, 1, 0)$ as a counter-example.

(5) Similarly, we observe that $0_{\mathbb{R}^3} \in E$, yet E is not a vector space. To verify this, the reader may consider the vectors $(1, 0, 1)$ and $(1, 0, -1)$ and check that their sum does not belong to E. This demonstrates that E is not closed under addition, confirming that it is not a subspace of $\mathbb{R}^3$.

(6) Although E involves squares, it is still a subspace over $\mathbb{R}$, since

$$x^2 + y^2 + z^2 = 0 \Longleftrightarrow x = y = z = 0.$$

Thus, E is simply $\{(0, 0, 0)\}$, which is trivially a vector space.

(7) For example, $(0, 1), (1, 0) \in E$. However,

$$(0, 1) + (1, 0) = (1, 1) \notin E,$$

because $1 \times 1 = 1 \neq 0$.

(8) First, $0_{\mathbb{R}^4} = (0, 0, 0, 0) \in E$, since $0 + 0 = 0$ and $0 + 2(0) = 0$ (at this stage, we cannot conclude anything yet). We check that E is a vector space. Let $(x, y, z, t), (x', y', z', t') \in E$, meaning

$$x + y = 0, \quad z + 2t = 0, \quad \text{and } x' + y' = 0, \quad z' + 2t' = 0.$$

Let $\alpha, \beta \in \mathbb{R}$. Then,

$$\alpha(x, y, z, t) + \beta(x', y', z', t')$$
$$= (\alpha x + \beta x', \alpha y + \beta y', \alpha z + \beta z', \alpha t + \beta t').$$

So,

$$\alpha x + \beta x' + \alpha y + \beta y' = \alpha(x + y) + \beta(x' + y') = 0,$$

and

$$\alpha z + \beta z' + 2(\alpha t + \beta t') = \alpha(z + 2t) + \beta(z' + 2t') = 0.$$

Thus, E is a subspace of $\mathbb{R}^4$.

(9) We have:

$$E = \{(x, y) \in \mathbb{R}^2 : x = 0\} = \{(0, y) : y \in \mathbb{R}\}$$
$$= \{y(0, 1) : y \in \mathbb{R}\},$$

showing that E is a vector space because it is spanned by one vector.

(10) We have:

$$E = \{(x, y, z) \in \mathbb{R}^3 : x = y = z\}$$
$$= \{(x, x, x) : x \in \mathbb{R}\}$$
$$= \{x(1, 1, 1) : x \in \mathbb{R}\}.$$

Thus, E is a linear subspace because it is spanned by one vector.

(11) Writing

$$E = \{(x, y, x) : x, y \in \mathbb{R}\} = \{x(1, 0, 1) + y(0, 1, 0) : x, y \in \mathbb{R}\}$$

shows that E is spanned by two vectors, which makes it a vector subspace.

(12) We may show that $E = \text{span}\{u, v, w\}$, with $u = (1, 0, 0, 0)$, $v = (0, 1, 1, 0)$, and $w = (0, 0, 0, 1)$. Thus, E is vector subspace of $\mathbb{R}^4$.

Exercise 2.3.10. (Cf. Exercise 2.4.5) Show that the only subspaces of $\mathbb{R}$ are $\mathbb{R}$ itself and $\{0\}$. What are the only subspaces of $\mathbb{R}^2$?

Solution 2.3.10. Let E be a subspace of $\mathbb{R}$. Since $\dim \mathbb{R} = 1$, $\dim E \leq 1$, meaning that $\dim E = 0$ or $\dim E = 1$. If $\dim E = 0$, then $E = \{0\}$. If $\dim E = 1$, given that E is a subspace of $\mathbb{R}$, it results in $E = \mathbb{R}$.

A similar approach can be used to handle subspaces of $\mathbb{R}^2$. Consider E as a subspace of $\mathbb{R}^2$. Then, the fact that $\dim E \leq 2$ signifies that $\dim E$ can be 0, 1, or 2. As above, $\dim E = 0$ gives $E = \{0\}$, while $\dim E = 2$ yields $E = \mathbb{R}^2$. The only slightly nontrivial case occurs when $\dim E = 1$, indicating that E is the span of a certain vector $u \in E$. In other words, $E = \{\alpha u : \alpha \in \mathbb{R}\}$.

Exercise 2.3.11. Determine which of the following sets form a subspace of $M_2(\mathbb{R})$:

(1) The set of 2×2 matrices with natural number entries.
(2) The set of 2×2 matrices with a zero determinant. (*Recall that the determinant, a notion yet to be defined, of a matrix* $\begin{pmatrix} a & b \\ c & d \end{pmatrix}$ *is given by the scalar* $ad - bc$.)
(3) The set of matrices of the form
$$\begin{pmatrix} a & b \\ 0 & c \end{pmatrix}.$$

Solution 2.3.11.

(1) No, this set is closed under addition but not under scalar multiplication. Indeed, the matrix
$$\begin{pmatrix} 1 & 2 \\ 1 & 1 \end{pmatrix}$$
belongs to the set (since $1, 2 \in \mathbb{N}$), but
$$\frac{1}{2} \begin{pmatrix} 1 & 2 \\ 1 & 1 \end{pmatrix} = \begin{pmatrix} 1/2 & 1 \\ 1/2 & 1/2 \end{pmatrix}$$
does not, as $1/2 \notin \mathbb{N}$. Thus, the given set is not a subspace.

(2) The answer is negative. For instance, consider the matrices

$$\begin{pmatrix} 3 & 6 \\ 1 & 2 \end{pmatrix} \quad \text{and} \quad \begin{pmatrix} 2 & 2 \\ 0 & 0 \end{pmatrix}.$$

Their determinants are both zero. However, the sum of these matrices is

$$\begin{pmatrix} 5 & 8 \\ 1 & 2 \end{pmatrix},$$

which has a nonzero determinant, namely 2. Hence, the set is not closed under addition and is not a subspace.

(3) The answer is positive. Set

$$E = \left\{ \begin{pmatrix} a & b \\ 0 & c \end{pmatrix} : a, b, c \in \mathbb{R} \right\}.$$

First, $0_{M_2(\mathbb{R})} \in E$. Consider two arbitrary matrices in E, i.e., $\begin{pmatrix} a & b \\ 0 & c \end{pmatrix}$ and $\begin{pmatrix} a' & b' \\ 0 & c' \end{pmatrix}$. Let α, β be arbitrary real scalars. Computing $\alpha \begin{pmatrix} a & b \\ 0 & c \end{pmatrix} + \beta \begin{pmatrix} a' & b' \\ 0 & c' \end{pmatrix}$ results in $\begin{pmatrix} \alpha a + \beta a' & \alpha b + \beta b' \\ 0 & \alpha c + \beta c' \end{pmatrix}$, which stays in E. Hence, the set is closed under addition and scalar multiplication, proving that E is a subspace of $M_2(\mathbb{R})$.

Exercise 2.3.12. Let $E = \mathcal{F}(\mathbb{R}, \mathbb{R})$ be the vector space of functions defined from $\mathbb{R}$ to $\mathbb{R}$. Identify among the following sets those that form a subspace of E:

(1) $A = \{f \in E : f(x) \leq 0, \text{ for all } x \in \mathbb{R}\}$,
(2) $B = \{f \in E : f(0) = 0\}$,
(3) $C = \{f \in E : f(0) = 2\}$,
(4) D is the set of constant functions.
(5) F is the set of differentiable functions.

Solution 2.3.12.

(1) The identity element of E is the zero function, denoted as $f(x) = 0$ for all $x \in \mathbb{R}$, which clearly belongs to A. For any f and g in A, it follows that $f(x) \leq 0$ and $g(x) \leq 0$ for all $x \in \mathbb{R}$. Thus, $f(x) + g(x) \leq 0$ for all $x \in \mathbb{R}$, which implies that $f + g \in A$. However, if $\lambda \in \mathbb{R}$ and $f \in A$, it is not necessarily true that $\lambda f \in A$. For example, if $\lambda = -1 \in \mathbb{R}$ and

$f(x) = -e^x$, then $f \in A$, while $\lambda f(x) = +e^x > 0$, indicating that $\lambda f \notin A$. In conclusion, A is not a subspace of E.

(2) Here, B is a subspace. Indeed, the null function belongs to B. To save a little time, we check directly that $\alpha f + \beta g \in B$ for all $\alpha, \beta \in \mathbb{R}$ and all $f, g \in B$. Let $\alpha, \beta \in \mathbb{R}$ and $f, g \in B$. Then, $f(0) = 0$ and $g(0) = 0$. So,

$$\alpha f(0) + \beta g(0) = \alpha \times 0 + \beta \times 0 = 0.$$

Thus, $\alpha f + \beta g \in B$.

(3) The set C is not a subspace because the zero vector, i.e., the zero function, does not belong to C.

(4) The set D is a subspace. Readers should be cautious regarding the identity element. D consists of *all* constant functions, so it definitely contains the zero function. Additionally, the sum of any two constant functions is still constant, and multiplying any constant function by a real scalar results in a constant function. This concludes the proof.

(5) The set F is a subspace of E because, based on standard calculus results, we know that the zero function is differentiable; the sum of two differentiable functions is differentiable, and a differentiable function multiplied by a scalar remains differentiable.

Exercise 2.3.13. Let $E = \mathbb{R}[X]$. Are the following sets subspaces of E:

(1) $A = \{P \in E : P'(0) = 1\}$,
(2) $B = \{P \in E : P(0) = P(1)\}$?

Solution 2.3.13.

(1) The identity element of E is the zero polynomial. However, it does not belong to A because if $P(x) = 0$ for all $x \in \mathbb{R}$, then its derivative satisfies $P'(x) = 0$ for all $x \in \mathbb{R}$, which in particular gives $P'(0) = 0 \neq 1$. Hence, A is not a subspace of E.

(2) The zero polynomial P_0 belongs to B since $P_0(0) = 0 = P_0(1)$. Next, let $P, Q \in B$ and $\alpha, \beta \in \mathbb{R}$. By the definition of B, we have $P(0) = P(1)$ and $Q(0) = Q(1)$. To show that $\alpha P + \beta Q \in B$, we need to check that

$$(\alpha P + \beta Q)(0) = (\alpha P + \beta Q)(1),$$

which follows directly from

$$\alpha P(0) + \beta Q(0) = \alpha P(1) + \beta Q(1).$$

Thus, B is a subspace of E.

Exercise 2.3.14.

(1) Express the vector $(1, -2, 5)$ as a linear combination of the vectors:
$$u = (1, 1, 1), \quad v = (1, 2, 3), \quad w = (2, -1, 1).$$

(2) Can we express the vector $(2, -5, 3)$ as a linear combination of the vectors:
$$u = (1, -3, 2), \quad v = (2, -4, -1), \quad w = (1, -5, 7)?$$

Solution 2.3.14.

(1) The goal is to find real scalars a, b, and c such that:
$$(1, -2, 5) = au + bv + cw = a(1, 1, 1) + b(1, 2, 3) + c(2, -1, 1),$$
i.e.,
$$(1, -2, 5) = (a, a, a) + (b, 2b, 3b) + (2c, -c, c),$$
i.e.,
$$(1, -2, 5) = (a + b + 2c, a + 2b - c, a + 3b + c).$$

Thus, the system to solve is:
$$\begin{cases} a + b + 2c = 1, \\ a + 2b - c = -2, \\ a + 3b + c = 5. \end{cases}$$

The solution to this system is given by:
$$a = -6, \quad b = 3, \quad c = 2.$$

Therefore, we have:
$$(1, -2, 5) = -6u + 3v + 2w.$$

(2) Let us now try to find real scalars a, b, and c such that:
$$(2, -5, 3) = a(1, -3, 2) + b(2, -4, -1) + c(1, -5, 7),$$
i.e.,
$$(2, -5, 3) = (a, -3a, 2a) + (2b, -4b, -b) + (c, -5c, 7c),$$
i.e.,
$$(2, -5, 3) = (a + 2b + c, -3a - 4b - 5c, 2a - b + 7c).$$

Thus, we are led to solve the system:
$$\begin{cases} a + 2b + c = 2, \\ -3a - 4b - 5c = -5, \\ 2a - b + 7c = 3. \end{cases}$$

From the first equation, we get $a = 2 - 2b - c$. Substituting this into the other two equations, we obtain: $-5b + 5c = -1$ and $2b - 2c = 1$, i.e., $b - c = 1/5$ and $b - c = 1/2$, which leads to a contradiction. Therefore, the system does not admit a solution, meaning that the vector $(2, -5, 3)$ cannot be written as a linear combination of u, v, and w.

Exercise 2.3.15. Do we have $\text{span}\{u_1, u_2, u_3\} = \mathbb{R}^3$ in the following cases:

(1) $u_1 = (1, 1, 1)$, $u_2 = (1, 1, 0)$, and $u_3 = (1, 0, 0)$,
(2) $u_1 = (1, 1, 2)$, $u_2 = (1, 0, 1)$, and $u_3 = (2, 1, 3)$?

Solution 2.3.15.

(1) Let $(x, y, z) \in \mathbb{R}^3$. We have to check if we can express (x, y, z) as a linear combination of u_1, u_2, u_3, i.e., if there exist scalars $a, b, c \in \mathbb{R}$ such that:

$$(x, y, z) = au_1 + bu_2 + cu_3.$$

We have:

$$(x, y, z) = a(1, 1, 1) + b(1, 1, 0) + c(1, 0, 0),$$

which simplifies to:

$$(x, y, z) = (a + b + c, a + b, a).$$

This gives the system of equations:

$$x = a + b + c,$$
$$y = a + b,$$
$$z = a.$$

From the third equation, $a = z$. Substituting this into the second equation $y = a + b$, we get:

$$y = z + b \implies b = y - z.$$

Substituting $a = z$ and $b = y - z$ into the first equation, we get:

$$x = z + (y - z) + c \implies x = y + c \implies c = x - y.$$

Thus, we can express (x, y, z) as:

$$(x, y, z) = zu_1 + (y - z)u_2 + (x - y)u_3.$$

Therefore,

$$\mathbb{R}^3 = \text{span}\{u_1, u_2, u_3\}.$$

(2) For $u_1 = (1, 1, 2), u_2 = (1, 0, 1), u_3 = (2, 1, 3)$, we seek to express $(x, y, z) \in \mathbb{R}^3$ as a linear combination of u_1, u_2, u_3. We need to solve the system:

$$(x, y, z) = au_1 + bu_2 + cu_3,$$

which leads to:

$$(x, y, z) = a(1, 1, 2) + b(1, 0, 1) + c(2, 1, 3),$$

simplifying to:

$$(x, y, z) = (a + b + 2c, a + c, 2a + b + 3c).$$

This gives the system of equations:

$$x = a + b + 2c,$$
$$y = a + c,$$
$$z = 2a + b + 3c.$$

Readers can check that the previous system is not solvable. Thus, the vectors u_1, u_2, u_3 do not span $\mathbb{R}^3$.

Exercise 2.3.16. Write the following polynomials as a linear combination of $p_1 = 2 + x + 4x^2$, $p_2 = 1 - x + 3x^2$, and $p_3 = 3 + 2x + 5x^2$:
  (1) $-9 - 7x - 15x^2$,
  (2) $6 + 11x + 6x^2$,
  (3) 0,
  (4) $7 + 8x + 9x^2$.

Solution 2.3.16. We will carry out the first case in detail. For the remaining cases, the reader can apply the same method, though we will provide the final results.

(1) The method consists of finding real scalars a, b, and c such that

$$-9 - 7x - 15x^2 = ap_1 + bp_2 + cp_3.$$

Expanding the right-hand side, we obtain

$$-9 - 7x - 15x^2$$
$$= a(2 + x + 4x^2) + b(1 - x + 3x^2) + c(3 + 2x + 5x^2),$$

which simplifies to

$$-9 - 7x - 15x^2$$
$$= (4a + 3b + 5c)x^2 + (a - b + 2c)x + (2a + b + 3c).$$

We present two methods to determine the real scalars a, b, and c.

(a) **First method:** Since the equation holds for all x, we substitute three values of x (chosen to yield a solvable system):

(i) $x = 0$ gives: $2a + b + 3c = -9$,
(ii) $x = 1$ gives: $7a + 3b + 10c = -31$,
(iii) $x = -1$ gives: $5a + 5b + 6c = -17$.

Solving this system yields $a = -2$, $b = 1$, and $c = -2$. Hence,

$$-9 - 7x - 15x^2 = -2p_1 + p_2 - 2p_3.$$

(b) **Second method:** Comparing coefficients, we obtain the system:

$$\begin{cases} 4a + 3b + 5c = -15, \\ a - b + 2c = -7, \\ 2a + b + 3c = -9. \end{cases}$$

Solving this system gives $a = -2$, $b = 1$, and $c = -2$, leading to the same result:

$$-9 - 7x - 15x^2 = -2p_1 + p_2 - 2p_3.$$

(2) We obtain:

$$6 + 11x + 6x^2 = 4p_1 - 5p_2 + p_3.$$

(3) The answer is:

$$0 = 0p_1 + 0p_2 + 0p_3.$$

(4) The solution is:

$$7 + 8x + 9x^2 = 0p_1 - 2p_2 + 3p_3.$$

Exercise 2.3.17. Let $f(x) = \cos^2 x$ and $g(x) = \sin^2 x$ (be defined from $\mathbb{R}$ to $\mathbb{R}$). Which of the following functions belong to the space spanned by f and g:

(1) $\cos 2x$,
(2) $1 + x^2$,
(3) 2,
(4) $\sin x$,
(5) 0.

Solution 2.3.17.

(1) $x \mapsto \cos 2x$ belongs to the space spanned by f and g because:

$$\cos 2x = 1 \cdot \cos^2 x - 1 \cdot \sin^2 x, \quad \forall x \in \mathbb{R}.$$

(2) The function $x \mapsto 1+x^2$ does not belong to the space spanned by f and g because we cannot find real scalars a and b such that

$$1 + x^2 = a \cos^2 x + b \sin^2 x.$$

Indeed, let us try to find a and b. Then,

$$x = 0 \Longrightarrow 1 + 0^2 = a \cos^2 0 + b \sin^2 0 \Longleftrightarrow a = 1$$

and

$$x = \pi \Longrightarrow 1 + \pi^2 = a \cos^2 \pi + b \sin^2 \pi \Longleftrightarrow 1 = 1 + \pi^2,$$

which is impossible.

Alternatively, assuming $1+x^2 = a \cos^2 x + b \sin^2 x$ for some $a, b \in \mathbb{R}^*$ and all x, we observe that differentiating the left-hand side three times results in 0, while differentiating the right-hand side the same number of times does not yield the zero function.

(3) The fundamental trigonometric identity ensures that the function $x \mapsto 2$ is a linear combination of f and g because:

$$2 = 2 \cos^2 x + 2 \sin^2 x, \quad \forall x \in \mathbb{R}.$$

(4) $x \mapsto \sin x$ does not belong to the space spanned by f and g because we cannot find real scalars a and b such that:

$$\sin x = a \cos^2 x + b \sin^2 x.$$

Indeed, if $\sin x = a \cos^2 x + b \sin^2 x$ for some a, b and all x, then it would hold for $x = \pi/4$ and $x = -\pi/4$ separately. However, the first case yields $a + b = \sqrt{2}$, and the second $a + b = -\sqrt{2}$. Since the system of equations leads to a contradiction, there is no pair of real numbers a and b that can satisfy the equation $\sin x = a \cos^2 x + b \sin^2 x$.

(5) The constant function equal to 0 on $\mathbb{R}$ belongs to the space spanned by f and g because:

$$0 = 0 \cos^2 x + 0 \sin^2 x, \quad \forall x \in \mathbb{R}.$$

In fact, 0 belongs to the span of any set!

Exercise 2.3.18. Are the following vectors linearly independent in E?

(1) $u = (1, 3)$ and $v = (2, 4)$ in $E = \mathbb{R}^2$.
(2) $u = (2, 3)$ and $v = (-4, -6)$ in $E = \mathbb{R}^2$.
(3) $u = (0, 1, 2)$ and $v = (1, 2, 0)$ in $E = \mathbb{R}^3$.
(4) $u = (2, -2, 2)$ and $v = (-3, 3, -3)$ in $E = \mathbb{R}^3$.

Solution 2.3.18. In general, in a vector space E, we can determine if the equation $au+bv = 0_E$ results in $a = b = 0$ or not to see whether or not u and v are linearly independent. When we have only two vectors, these are linearly dependent if and only if one is a multiple of the other. Thus,

- (1) u and v are linearly independent.
- (2) u and v are linearly dependent because $v = -2u$.
- (3) u and v are linearly independent.
- (4) u and v are linearly dependent for

$$-3u = 2v, \text{ i.e. } u = -\frac{2}{3}v \left(\text{or } v = -\frac{3}{2}u \right).$$

Exercise 2.3.19. Are the following vectors linearly independent in $\mathbb{R}^3$?

- (1) $u_1 = (1,2,3)$, $u_2 = (4,5,6)$, $u_3 = (7,8,9)$, and $u_4 = (10,11,12)$
- (2) $u = (1,2,5)$, $v = (2,5,1)$, and $w = (1,5,2)$.
- (3) $u = (0,0,0)$, $v = (1,2,4)$, and $w = (1,3,7)$.
- (4) $u = (1,2,-1)$, $v = (1,0,1)$, and $w = (-1,2,-3)$.

Solution 2.3.19.

- (1) Without performing calculations, since $\dim \mathbb{R}^3 = 3$ and we have *four* vectors, we can deduce that (u_1, u_2, u_3, u_4) must be linearly *dependent*!
- (2) The three given vectors are linearly independent. Indeed, let $a, b, c \in \mathbb{R}$ be such that

$$au + bv + cw = (0,0,0).$$

Then,

$$(a, 2a, 5a) + (2b, 5b, b) + (c, 5c, 2c) = (0,0,0),$$

that is,

$$(a + 2b + c, 2a + 5b + 5c, 5a + b + 2c) = (0,0,0).$$

So, the system to solve is:

$$\begin{cases} a + 2b + c = 0, \\ 2a + 5b + 5c = 0, \\ 5a + b + 2c = 0. \end{cases}$$

From the first equation, $a = -c - 2b$, and substituting it into the other equations, we get $3c+b = 0$ and $-3c-9b = 0$. Adding these two equations together yields $b = 0$. Consequently, $c = 0$, and thus $a = 0$. Therefore, the vectors u, v, and w are linearly independent.

(3) Since one of the vectors, more precisely, u, is the zero vector, we conclude that u, v, and w are linearly dependent.

(4) We will determine whether the vectors are linearly independent or not. Let a, b, and c be three real scalars such that

$$au + bv + cw = (0,0,0).$$

Then,

$$(a + b - c, 2a + 2c, -a + b - 3c) = (0,0,0),$$

i.e.,

$$\begin{cases} a + b - c = 0, \\ 2a + 2c = 0, \\ -a + b - 3c = 0. \end{cases}$$

So,

$$\begin{cases} a + b - c = 0, \\ 2a + 2c = 0, \\ 2b - 4c = 0. \end{cases}$$

We have infinitely many solutions where $a = -c$, $b = 2c$, and $c \in \mathbb{R}$. For instance, if $c = 1$, then $b = 2$ and $a = -1$, which gives the following linear combination:

$$-(1, 2, -1) + 2(1, 0, 1) + (-1, 2, -3) = (0, 0, 0),$$

which then confirms that the vectors are linearly dependent.

Exercise 2.3.20. Prove that the vectors $(1+i, 2i)$ and $(1, 1+i)$ are linearly *dependent* on $\mathbb{C}^2$ as a vector space over $\mathbb{C}$, yet they are linearly *independent* on $\mathbb{C}^2$ as a vector space over $\mathbb{R}$.

Solution 2.3.20. We notice that:

$$(1 + i)(1, 1 + i) = (1 + i, (1 + i)^2) = (1 + i, 2i).$$

Then, the vectors are linearly dependent on $\mathbb{C}^2$ as a $\mathbb{C}$-vector space.

Let a and b be real scalars such that $a(1+i, 2i) + b(1, 1+i) = (0, 0)$. Then,

$$(a + ia, 2ia) + (b, b + ib) = (0, 0) \iff (a + b + i, b + ib + 2ia) = (0, 0).$$

Hence,

$$\begin{cases} a + b = 0, \\ a = 0, \\ b = 0, \\ 2a + b = 0. \end{cases}$$

In other words, $a = b = 0$, and the vectors are linearly independent on $\mathbb{C}^2$ as an $\mathbb{R}$-vector space.

Exercise 2.3.21. Are the following sets of functions from $\mathbb{R}$ to $\mathbb{R}$ linearly independent:

(1) $\{1, X, X^2, (X-3)(X+1)\}$ (in $\mathbb{R}_2[X]$),
(2) $\{\sin(1+x), \sin(2+x), \sin(3+x)\}$,
(3) $\{x, \sin x, \cos x\}$?

Solution 2.3.21.

(1) Since the cardinality of the set $\{1, X, X^2, (X-3)(X+1)\}$ is 4, which is greater than the dimension of $\mathbb{R}_2[X]$ (which is 3), the elements of the set $\{1, X, X^2, (X-3)(X+1)\}$ are necessarily linearly dependent.

For anyone who missed it, here's how they might present their arguments:

$$(X-3)(X+1) = X^2 - 2X - 3 = -3 \times \mathbf{1} - 2 \times \mathbf{X} + 1 \times \mathbf{X^2}.$$

(2) No, indeed, we can prove using well-known trigonometric formulas that

$$\sin(1+x) + \sin(3+x) = 2\sin(2+x)\cos(1) = \alpha\sin(2+x),$$

where $\alpha = 2\cos(1)$ is a real scalar.

(3) Let $a, b, c \in \mathbb{R}$ be such that

$$ax + b\sin(x) + c\cos(x) = 0, \quad \forall x \in \mathbb{R}.$$

To solve for the unknowns, we can choose specific values for x to obtain a system of three equations. Setting $x = 0$, we get $c = 0$. Setting $x = \frac{\pi}{2}$, we get $\frac{a\pi}{2} + b = 0$. Finally, setting $x = \pi$, we get $a\pi - c = 0$. So, from the first equation, we find $c = 0$. From the third equation, we find $a = 0$. From the second equation, we find $b = 0$. Therefore, the set $\{x, \sin(x), \cos(x)\}$ is linearly independent.

Exercise 2.3.22. Let the functions (f_n) $(n = 1, 2, 3)$ be defined from $\mathbb{R}$ to $\mathbb{R}$. Determine whether the following sets are free:

(1) $f_1(x) = e^x$, $f_2(x) = e^{2x}$, and $f_3(x) = e^{3x}$.
(2) $f_1(x) = e^x$, $f_2(x) = e^{-x}$, and $f_3(x) = \cosh x$.
(3) $f_1(x) = |x-1|$, $f_2(x) = |x-2|$, and $f_3(x) = |x-3|$.

Solution 2.3.22.

(1) The vectors are linearly independent. Let $a, b, c \in \mathbb{R}$ be such that

$$ae^x + be^{2x} + ce^{3x} = 0, \quad \forall x \in \mathbb{R}.$$

There are various methods to reach $a = b = c = 0$. For instance, we can evaluate at specific values to x, then form a system with three unknowns to solve. Alternatively, we divide everything by e^{3x}, which yields

$$ae^{-2x} + be^{-x} + c = 0, \ \forall x \in \mathbb{R}.$$

As we let x tend to $+\infty$, we find that $c = 0$. Returning to the first equation, it becomes $ae^x + be^{2x} = 0$. Multiplying by e^{-2x}, we get

$$ae^{-x} + b = 0, \ \forall x \in \mathbb{R}.$$

As we let x tend to $+\infty$, we find $b = 0$. Thus, $ae^x = 0$ for all x. Since $e^x \neq 0$, for all $x \in \mathbb{R}$, we find $a = 0$.

(2) Recall that $\cosh x = \frac{e^x + e^{-x}}{2}$, which implies $f_3 = \frac{1}{2}f_1 + \frac{1}{2}f_2$, making the three vectors linearly dependent.

(3) The given vectors are linearly independent. Let $a, b, c \in \mathbb{R}$ be such that

$$a|x - 1| + b|x - 2| + c|x - 3| = 0$$

for all real x.

Among the methods we can use, we have: The "zero" on the right is a differentiable function for all x, especially at $x = 1$, whereas $x \mapsto |x - 1|$ is not differentiable at $x = 1$ unless $a = 0$.

If $a = 0$, the equation becomes $b|x - 2| + c|x - 3| = 0$. The "zero" on the right is a differentiable function for all x, especially at $x = 2$, whereas $x \mapsto |x - 2|$ is not differentiable at $x = 2$ unless $b = 0$.

If $b = 0$, the equation becomes $c|x - 3| = 0$, which holds for all x only if $c = 0$.

Thus, $a = b = c = 0$, and the given vectors are linearly independent.

Exercise 2.3.23. Answer the same question as in the previous exercise:

(1) $\{\ln x, \ln^2 x, \ln^3 x\}$ defined on $(0, \infty)$,
(2) $\left\{\frac{1}{x-2}, \frac{1}{x+2}, \frac{2x}{x^2-4}\right\}$ defined on $(-2, 2)$.

Solution 2.3.23.

(1) The vectors are linearly independent! Let $a, b, c \in \mathbb{R}$ such that

$$a \ln x + b \ln^2 x + c \ln^3 x = 0.$$

When $x = e$, then $x = e^2$, and $x = e^3$, we will obtain a system of equations whose only solution is $a = b = c = 0$.

(2) The given vectors are linearly dependent. Indeed, since $x^2 - 4 = (x-2)(x+2)$, we can write:

$$\frac{2x}{x^2 - 4} = \frac{2x}{(x-2)(x+2)} = \frac{\alpha}{x-2} + \frac{\beta}{x+2}.$$

Let us find the values of α and β. Multiplying the equation above by $(x-2)$, we get:

$$\frac{2x}{x+2} = \alpha + \frac{\beta(x-2)}{x+2}$$

and when we let x approach 2, we find: $\alpha = \frac{2 \times 2}{2+2} = 1$.

To find β, we go back to the first equation and multiply by $(x+2)$, we find:

$$\frac{2x}{x-2} = \frac{\alpha(x+2)}{x-2} + \beta$$

and we let x tend to -2 to get $\beta = \frac{2 \times (-2)}{-2-2} = 1$. Thus,

$$\frac{2x}{x^2 - 4} = \frac{1}{x-2} + \frac{1}{x+2},$$

as required.

Exercise 2.3.24. Show that the *finite* set $\{\cos x, \cos 2x, \ldots, \cos nx\}$ is free.

Solution 2.3.24. We prove this using the principle of induction. Assume the elements of P_n: $\{\cos x, \cos 2x, \ldots, \cos nx\}$ are linearly independent.

The base case is true because $\{\cos x\}$ is not the null function.

Suppose the elements of $\{\cos x, \ldots, \cos nx\}$ are linearly independent, and let us show that $\{\cos x, \ldots, \cos nx, \cos(n+1)x\}$ is a free set. Let $a_1, \ldots, a_n, a_{n+1}$ be real scalars such that

$$a_1 \cos x + \cdots + a_n \cos nx + a_{n+1} \cos(n+1)x = 0.$$

Differentiating the previous equation two times (with respect to x) gives

$$-a_1 \cos x - \cdots - a_n n^2 \cos nx - a_{n+1}(n+1)^2 \cos(n+1)x = 0.$$

Adding the previous equation with the one before it, multiplied by $(n+1)^2$, implies that

$$((n+1)^2 - 1)a_1 \cos x + \cdots + ((n+1)^2 - n^2)a_n n^2 \cos nx = 0.$$

It is a linear combination of the vectors $\{\cos x, \cos 2x, \ldots, \cos nx\}$, which are assumed to be linearly independent. So, all the coefficients must vanish, meaning that

$$((n+1)^2 - k^2)a_k = 0, \quad \forall k = 1, \ldots, n.$$

Hence, $a_1 = a_2 = \cdots = a_n = 0$. Thus, $a_{n+1} \cos(n+1)x = 0$, which leads to $a_{n+1} = 0$. Consequently, the vectors of the family $\{\cos x, \ldots, \cos nx, \cos(n+1)x\}$ are linearly independent.

Exercise 2.3.25. Reconsider the question posed in the previous exercise with the *infinite* set $\{e^{nx}\}_{n\in\mathbb{N}} = \{1, e^x, \ldots, e^{nx}, \ldots\}$.

Solution 2.3.25. The elements are linearly independent! To demonstrate this, we need to show that the elements of any finite subset of the given family are linearly independent. A finite subset is of the form: $\{e^{a_0 x}, e^{a_1 x}, \ldots, e^{a_m x}\}$, where a_i, $0 \le i \le m$ are natural integers. We can choose a_i in ascending order, so we suppose that: $a_0 < a_1 < \cdots < a_m$. Let $b_0, b_1, \ldots, b_m$ be real scalars such that

$$b_0 e^{a_0 x} + b_1 e^{a_1 x} + \cdots + b_m e^{a_m x} = 0$$

for all $x \in \mathbb{R}$. Then,

$$e^{a_m x}(b_0 e^{(a_0 - a_m)x} + b_1 e^{(a_1 - a_m)x} + \cdots + b_m) = 0,$$

still for all $x \in \mathbb{R}$. Letting x tend to $+\infty$ yields $b_m = 0$. The same approach can be applied to demonstrate that all the other coefficients also vanish, showing that the elements of $\{e^{a_0 x}, e^{a_1 x}, \ldots, e^{a_m x}\}$ are linearly independent.

Exercise 2.3.26. Consider $\mathbb{R}$ as a vector space over $\mathbb{Q}$.
(1) Show that 1, $\sqrt{2}$, and $\sqrt{3}$ are pairwise linearly independent.
(2) Show that 1, $\sqrt{2}$, and $\sqrt{3}$ are linearly independent.

Solution 2.3.26. First, recall that if p is not a perfect square (i.e. it is not of the form $p = s^2$ for any integer s), then $\sqrt{p}$ is irrational.

(1) Let us show that (1 and $\sqrt{2}$), (1 and $\sqrt{3}$), ($\sqrt{2}$ and $\sqrt{3}$) are linearly independent.

Let $a, b \in \mathbb{Q}$ be such that $a \times 1 + b\sqrt{2} = 0$. If $b = 0$, then $a = 0$. If $b \ne 0$, then we have $\sqrt{2} = -\frac{a}{b}$. But $\sqrt{2}$ is not rational and $-\frac{a}{b}$ is rational. Therefore, $a = b = 0$.

Let $a, b \in \mathbb{Q}$ such that $a \times 1 + b\sqrt{3} = 0$. If $b = 0$, then $a = 0$. If $b \ne 0$, then we will have $\sqrt{3} = -\frac{a}{b}$. However, $\sqrt{3}$ is not rational, and $-\frac{a}{b}$ is rational. Therefore, $a = b = 0$.

Let $a, b \in \mathbb{Q}$ be such that $a\sqrt{2} + b\sqrt{3} = 0$. If $b = 0$, then $a = 0$. If $b \neq 0$, then we have $\frac{\sqrt{3}}{\sqrt{2}} = -\frac{a}{b}$. But $\frac{\sqrt{3}}{\sqrt{2}}$ is not rational, while $-\frac{a}{b}$ is rational. So, we conclude that $a = b = 0$.

(2) Let $a, b, c \in \mathbb{Q}$ be such that $a + b\sqrt{2} + c\sqrt{3} = 0$. Then, we have:

$$b\sqrt{2} + c\sqrt{3} = -a \implies (b\sqrt{2} + c\sqrt{3})^2 \in \mathbb{Q}.$$

So, $(b\sqrt{2} + c\sqrt{3})^2 = 2b^2 + 3c^2 + 2bc\sqrt{6} \in \mathbb{Q}$, that is, $bc\sqrt{6} \in \mathbb{Q}$. Since $\sqrt{6} \notin \mathbb{Q}$, we conclude that $bc = 0$. If only one of b or c were zero, then we would have either $\sqrt{3}$ or $\sqrt{2}$ in the set of rational numbers, which is impossible. Thus, $b = c = 0$, and it follows that $a = 0$ as well.

Exercise 2.3.27. Consider the following three vectors in $\mathbb{R}^3$:

$$u = (1, 0, -1), \quad v = (2, 0, -2), \quad \text{and} \quad w = (0, 1, 1).$$

(1) Is there any vector z in $\mathbb{R}^3$ such that u, v, and z are linearly independent?

(2) Reconsider the same question with w instead of v.

Solution 2.3.27.

(1) Observe that $v = 2u$, which makes u, v, and z linearly dependent!

(2) Since vector u is not a multiple of vector w, we can conclude that u and w are linearly independent. Let us choose a vector z such that u, w, and z are linearly independent. For example, let us take $z = (1, 0, 0)$.

To verify that u, w, and z are linearly independent, we can assume that $au + bw + cz = 0$ and solve for a, b, and c. This gives us the equation $(a + c, b, -a + b) = (0, 0, 0)$. By solving this system of equations, we find that $a = b = c = 0$, confirming that the vectors u, w, and z are indeed linearly independent.

Exercise 2.3.28. Determine the coordinates of the vector u with respect to the vectors of the *basis B* in E:

(1) $u = (4, -3)$, B is the standard basis of $E = \mathbb{R}^2$.

(2) $u = (4, -3)$, $B = \{(1, 1), (2, 3)\}$ in $E = \mathbb{R}^2$.

(3) $u = (x, y, z)$, B is the standard basis of $E = \mathbb{R}^3$.

(4) $u = (x, y, z)$, $B = \{(1, 1, 1), (1, 1, 0), (1, 0, 0)\}$ in $E = \mathbb{R}^3$.

Solution 2.3.28.

(1) The vector $u = (4, -3)$ is, by default, written with respect to the standard basis $\{e_1, e_2\}$. In other words,

$$(4, -3) = 4(1, 0) - 3(0, 1) = 4e_1 - 3e_2.$$

(2) To find the coordinates of the point $u = (4, -3)$ with respect to the basis B, we need to find the real scalars a and b that satisfy the equation $(4, -3) = a(1, 1) + b(2, 3)$. This gives us the system of equations:

$$\begin{cases} a + 2b = 4, \\ a + 3b = -3. \end{cases}$$

Solving this system, we find that $a = 18$ and $b = -7$. Therefore, the coordinates of $u = (4, -3)$ in the basis B are $(18, -7)$.

(3) In $\mathbb{R}^3$, a vector (x, y, z)'s coordinates with respect to the standard basis are simply the components of the vector itself. Therefore, the coordinates of (x, y, z) with respect to the standard basis $\{e_1, e_2, e_3\}$ in $\mathbb{R}^3$ are x, y, and z themselves, written as

$$(x, y, z) = x(1, 0, 0) + y(0, 1, 0) + z(0, 0, 1) = xe_1 + ye_2 + ze_3.$$

(4) Write

$$(x, y, z) = a(1, 1, 1) + b(1, 1, 0) + c(1, 0, 0),$$

where we need to find a, b, and c. This leads to the following system

$$\begin{cases} a + b + c = x, \\ a + b = y, \\ a = z. \end{cases}$$

We already have $a = z$, and from the second equation, we get $b = y - a = y - z$, so $c = x - b - a = x - y$. The new coordinates of u are: $(z, y - z, x - y)$.

Exercise 2.3.29. Over $\mathbb{R}^4$, determine whether the following set of vectors can be extended to form a basis of $\mathbb{R}^4$:

(1) $u = (1, 1, 1, 0)$, $v = (1, 0, 1, 2)$, and $w = (0, 1, 0, -2)$.
(2) $u = (1, -2, 5, -3)$, $v = (1, 0, 0, 1)$, and $w = (1, 1, -3, 2)$.
(3) $u = (1, -2, 5, 3)$ and $v = (0, 7, -9, 2)$.

Solution 2.3.29. To complete a basis in a vector space, we start with a set of linearly independent vectors and add additional vectors if needed, until the set spans the entire space. The resulting set of vectors will then form a basis for the vector space, provided that they remain linearly independent.

(1) One can notice that $u = w + v$. So, the vectors u, v, and w are linearly dependent, making it impossible for the set $\{u, v, w\}$ to be extended to form a basis of $\mathbb{R}^4$.

(2) We can quickly check that u, v, and w are linearly independent, which we leave to readers. Let us extend this family of vectors to form a basis of $\mathbb{R}^4$. We can accomplish this by adding $e_4 = (0, 0, 0, 1)$ to our set. Since the cardinality of $\{u, v, w, e_4\}$ is 4, which matches the dimension of $\mathbb{R}^4$, this set will be a basis if we can show that its elements are linearly independent. Let a, b, c, and d be such that

$$au + bv + cw + de_4 = (0, 0, 0, 0).$$

The equations derived from the previous equation are:

$$\begin{cases} a + b + c = 0, \\ -2a + c = 0, \\ 5a - 3c = 0, \\ -3a + b + 2c + d = 0. \end{cases}$$

By solving these equations, we find that $a = c = 0$ from the second and third equations, $b = 0$ from the first equation, and $d = 0$ from the last equation. This shows that the vectors are linearly independent and form a basis of $\mathbb{R}^4$.

(3) The vectors u and v are not proportional, so they are linearly independent. We can expand this set to form a basis of $\mathbb{R}^4$. It is easy to verify that the vectors in the set $\{u, v, e_3, e_4\}$ form a basis of $\mathbb{R}^4$ (where $e_3 = (0, 0, 1, 0)$ and $e_4 = (0, 0, 0, 1)$). This verification is left to the readers.

Exercise 2.3.30. Let $F = \{(x, y) \in \mathbb{R}^2 : x + y = 0\}$. Prove that F is a subspace of $\mathbb{R}^2$ and determine its dimension.

Solution 2.3.30. First, observe that $(0, 0) \in F$ since $(0, 0) \in \mathbb{R}^2$ and $0 + 0 = 0$. Let $(x, y), (x', y') \in F$ (so $x + y = 0$ and $x' + y' = 0$) and let $\alpha, \beta \in \mathbb{R}$. We need to show that $\alpha(x, y) + \beta(x', y') \in F$. Clearly, we have $\alpha(x, y) + \beta(x', y') \in \mathbb{R}^2$. Also, we have

$$\alpha(x, y) + \beta(x', y') = (\alpha x, \alpha y) + (\beta x', \beta y') = (\alpha x + \beta x', \alpha y + \beta y')$$

and

$$(\alpha x + \beta x', \alpha y + \beta y') \in F \iff \alpha x + \beta x' + \alpha y + \beta y' = 0.$$

However,

$$\alpha x + \beta x' + \alpha y + \beta y' = \alpha \underbrace{(x + y)}_{=0} + \beta \underbrace{(x' + y')}_{=0} = 0.$$

Thus, F is a subspace of $\mathbb{R}^2$.

To find the dimension of F, it is primordial to determine a basis of F. Writing

$$F = \{(x, y) \in \mathbb{R}^2 : y = -x\} = \{(x, -x) : x \in \mathbb{R}\}$$

leads to $F = \{x(1, -1) : x \in \mathbb{R}\}$. Hence, F is spanned by the (nonzero) vector $u = (1, -1)$, which makes $\{u\}$ a basis of F. Accordingly,

$$\dim F = \mathrm{card}\{u\} = 1.$$

Exercise 2.3.31. Find the dimension of the *subspaces* F of E in the following cases:
  (1) $F = \{(x, y, z) \in \mathbb{R}^3 : x + y - z = 0\}$, $E = \mathbb{R}^3$.
  (2) $F = \{(x, y, z, t) \in \mathbb{R}^4 : x - t = 0\}$, $E = \mathbb{R}^4$.

Solution 2.3.31.

(1) We can write

$$F = \{(x, y, z) \in \mathbb{R}^3 : x + y - z = 0\} = \{(x, y, z) \in \mathbb{R}^3 : z = x + y\},$$

that is,

$$F = \{(x, y, x + y) : x, y \in \mathbb{R}\} = \{(x, 0, x) + (0, y, y) : x, y \in \mathbb{R}\},$$

i.e.,

$$F = \{x \underbrace{(1, 0, 1)}_{u} + y \underbrace{(0, 1, 1)}_{v} : x, y \in \mathbb{R}\}.$$

So, F is spanned by u and v, that is, $F = \mathrm{span}\{u, v\}$. Since they are not proportional, they are linearly independent. Thus, $\{u, v\}$ is a basis of F, whereby $\dim F = 2$.

(2) We have

$$F = \{(x, y, z, t) \in \mathbb{R}^4 : x - t = 0\} = \{(x, y, z, x) : x, y, z \in \mathbb{R}\}.$$

So,

$$F = \{x \underbrace{(1, 0, 0, 1)}_{u} + y \underbrace{(0, 1, 0, 0)}_{v} + z \underbrace{(0, 0, 1, 0)}_{w} : x, y, z \in \mathbb{R}\}.$$

Hence, $F = \text{span}\{u, v, w\}$. Therefore, $\dim F = 3$ because u, v, and w are linearly independent.

Exercise 2.3.32. Show that

$$\underbrace{\text{span}\{(3,7,0),(5,0,-7)\}}_{:=E} = \underbrace{\text{span}\{(1,-1,-2),(2,3,-1)\}}_{:=F}.$$

Solution 2.3.32. Since E is spanned by two vectors, it forms a vector space of dimension at most 2. The same applies to F. Given that $(3,7,0)$ and $(5,0,-7)$ are linearly independent, we conclude that $\dim E = 2$. Similarly, $\dim F = 2$.

To prove $E = F$, it suffices to show that $E \subseteq F$ (or equivalently, $F \subseteq E$, since both have the same dimension).

Let $(x, y, z) \in E$. Then there exist scalars $\alpha, \beta \in \mathbb{R}$ such that

$$(x, y, z) = \alpha(3,7,0) + \beta(5,0,-7) = (3\alpha + 5\beta, 7\alpha, -7\beta).$$

We need to express (x, y, z) as a linear combination of the basis of F,

$$(x, y, z) = a(1,-1,-2) + b(2,3,-1) = (a + 2b, -a + 3b, -2a - b).$$

Equating components gives the system:

$$\begin{cases} 3\alpha + 5\beta = a + 2b, \\ 7\alpha = -a + 3b, \\ -7\beta = -2a - b. \end{cases}$$

Solving for a and b, we obtain:

$$a = 3\beta - \alpha, \quad b = 2\alpha + \beta.$$

Since these expressions yield real values for all $\alpha, \beta \in \mathbb{R}$, we conclude that $(x, y, z) \in F$, proving $E \subseteq F$.

Since $\dim E = \dim F = 2$, this implies $E = F$, completing the proof.

Exercise 2.3.33. Let E be the set of 2×3 matrices of the form:

$$\begin{pmatrix} a & b & 2b + c \\ b & 2c & c - a \end{pmatrix},$$

where $a, b, c \in \mathbb{R}$.

   (1) Provide two specific matrices that belong to E and another matrix that does not belong to E.

   (2) Show that E is a vector space.

   (3) Determine a basis of E.

Solution 2.3.33.

(1) We have:

$$\begin{pmatrix} 0 & 0 & 0 \\ 0 & 0 & 0 \end{pmatrix} = \begin{pmatrix} 0 & 0 & 2 \times 0 + 0 \\ 0 & 2 \times 0 & 0 - 0 \end{pmatrix} \in E.$$

Another example is $\begin{pmatrix} 1 & 2 & 4 \\ 2 & 0 & -1 \end{pmatrix}$, which corresponds to $a = 1$, $b = 2$, and $c = 0$.

The matrix $\begin{pmatrix} 1 & 1 & 0 \\ 2 & -8 & -3 \end{pmatrix}$ does not belong to E. One reason is that the second row's first entry, b, cannot simultaneously be 1 and 2. Specifically, the matrix requires that $b = 1$ based on the first row, but also $b = 2$ based on the second row. This creates a contradiction, meaning the matrix cannot be written in the form prescribed for matrices in E.

(2) Since $M_{2,3}(\mathbb{R})$ is a vector space, we need only show that E is a subspace (of $M_{2,3}(\mathbb{R})$). First, we observe that the identity element of $M_{2,3}(\mathbb{R})$, which is nothing but the matrix $\begin{pmatrix} 0 & 0 & 0 \\ 0 & 0 & 0 \end{pmatrix}$, does belong to E. Now, let $\alpha, \beta \in \mathbb{R}$. Let $A, B \in E$, so

$$A = \begin{pmatrix} a & b & 2b + c \\ b & 2c & c - a \end{pmatrix} \text{ and } B = \begin{pmatrix} a' & b' & 2b' + c' \\ b' & 2c' & c' - a' \end{pmatrix}, \text{ where}$$

$a, a', b, b', c, c' \in \mathbb{R}$. We have

$$\alpha A + \beta B = \alpha \begin{pmatrix} a & b & 2b + c \\ b & 2c & c - a \end{pmatrix} + \beta \begin{pmatrix} a' & b' & 2b' + c' \\ b' & 2c' & c' - a' \end{pmatrix}$$

$$= \begin{pmatrix} \alpha a + \beta a' & \alpha b + \beta b' & 2(\alpha b + \beta b') + \alpha c + \beta c' \\ \alpha b + \beta b' & 2(\alpha c + \beta c') & \alpha c + \beta c' - (\alpha a + \beta a') \end{pmatrix}$$

and the latter is in the form of matrices that belong to E, i.e., $\alpha A + \beta B \in E$, i.e. E is a vector space over $\mathbb{R}$.

(3) First, observe that $\dim E \leq \dim M_{2,3}(\mathbb{R}) = 6$. Now, we find a basis of E. Let $a, b, c \in \mathbb{R}$. We can write

$$\begin{pmatrix} a & b & 2b + c \\ b & 2c & c - a \end{pmatrix} = \begin{pmatrix} a & 0 & 0 \\ 0 & 0 & -a \end{pmatrix} + \begin{pmatrix} 0 & b & 2b \\ b & 0 & 0 \end{pmatrix}$$

$$+ \begin{pmatrix} 0 & 0 & c \\ 0 & 2c & c \end{pmatrix}$$

$$= a \begin{pmatrix} 1 & 0 & 0 \\ 0 & 0 & -1 \end{pmatrix} + b \begin{pmatrix} 0 & 1 & 2 \\ 1 & 0 & 0 \end{pmatrix}$$
$$\underbrace{\phantom{a \begin{pmatrix} 1 & 0 & 0 \\ 0 & 0 & -1 \end{pmatrix}}}_{M_1} \quad \underbrace{\phantom{b \begin{pmatrix} 0 & 1 & 2 \\ 1 & 0 & 0 \end{pmatrix}}}_{M_2}$$

$$+ c \underbrace{\begin{pmatrix} 0 & 0 & 1 \\ 0 & 2 & 1 \end{pmatrix}}_{M_3}.$$

So, E is spanned by $\{M_1, M_2, M_3\}$. To show that its elements are linearly independent (hence they form a basis), we need to demonstrate that $\alpha M_1 + \beta M_2 + \gamma M_3 = 0_{M_{2,3}(\mathbb{R})}$ only when $\alpha = \beta = \gamma = 0$. The equations

$$\alpha \begin{pmatrix} 1 & 0 & 0 \\ 0 & 0 & -1 \end{pmatrix} + \beta \begin{pmatrix} 0 & 1 & 2 \\ 1 & 0 & 0 \end{pmatrix} + \gamma \begin{pmatrix} 0 & 0 & 1 \\ 0 & 2 & 1 \end{pmatrix} = \begin{pmatrix} 0 & 0 & 0 \\ 0 & 0 & 0 \end{pmatrix}$$

simplify to

$$\begin{pmatrix} \alpha & \beta & 2\beta + \gamma \\ \beta & 2\gamma & \gamma - \alpha \end{pmatrix} = \begin{pmatrix} 0 & 0 & 0 \\ 0 & 0 & 0 \end{pmatrix},$$

which gives the following system:

$$\begin{cases} \alpha = 0, \\ \beta = 0, \\ 2\beta + \gamma = 0, \\ \beta = 0, \\ 2\gamma = 0, \\ \gamma - \alpha = 0. \end{cases}$$

So, $\alpha = \beta = \gamma = 0$ is the only solution that satisfies *all* equations. Thus, $\{M_1, M_2, M_3\}$ is a basis of E, whereby

$$\dim E = \mathrm{card}\{M_1, M_2, M_3\} = 3.$$

Exercise 2.3.34. Let

$$E = \left\{ A \in M_3(\mathbb{R}) : A = \begin{pmatrix} a & 0 & c \\ 0 & b & 0 \\ c & 0 & a \end{pmatrix}, \ a, b, c \in \mathbb{R} \right\}.$$

(1) Prove that E is a subspace of $M_3(\mathbb{R})$.
(2) Determine $\dim E$.

Solution 2.3.34.

(1) Since

$$\begin{pmatrix} a & 0 & c \\ 0 & b & 0 \\ c & 0 & a \end{pmatrix} = \begin{pmatrix} 0 & 0 & 0 \\ 0 & 0 & 0 \\ 0 & 0 & 0 \end{pmatrix} = 0_{M_3(\mathbb{R})},$$

for some $a = b = c = 0 \in \mathbb{R}$, it follows that $0_{M_3(\mathbb{R})} \in E$.
Next, let $\alpha, \beta \in \mathbb{R}$ and $A, B \in E$, and so

$$A = \begin{pmatrix} a & 0 & c \\ 0 & b & 0 \\ c & 0 & a \end{pmatrix} \text{ and } B = \begin{pmatrix} a' & 0 & c' \\ 0 & b' & 0 \\ c' & 0 & a' \end{pmatrix},$$

with $a, b, c; a', b', c' \in \mathbb{R}$. We have

$$\alpha A + \beta B = \alpha \begin{pmatrix} a & 0 & c \\ 0 & b & 0 \\ c & 0 & a \end{pmatrix} + \beta \begin{pmatrix} a' & 0 & c' \\ 0 & b' & 0 \\ c' & 0 & a' \end{pmatrix}$$

$$= \begin{pmatrix} \alpha a & 0 & \alpha c \\ 0 & \alpha b & 0 \\ \alpha c & 0 & \alpha a \end{pmatrix} + \begin{pmatrix} \beta a' & 0 & \beta c' \\ 0 & \beta b' & 0 \\ \beta c' & 0 & \beta a' \end{pmatrix}$$

$$= \begin{pmatrix} \alpha a + \beta a' & 0 & \alpha c + \beta c' \\ 0 & \alpha b + \beta b' & 0 \\ \alpha c + \beta c' & 0 & \alpha a + \beta a' \end{pmatrix} \in E.$$

(2) Recall that $\dim M_3(\mathbb{R}) = 3^2 = 9$, so necessarily, $\dim E \leq 9$. To find the dimension of E, we must determine a basis of it. Let $a, b, c \in \mathbb{R}$. We can write

$$\begin{pmatrix} a & 0 & c \\ 0 & b & 0 \\ c & 0 & a \end{pmatrix} = \begin{pmatrix} a & 0 & 0 \\ 0 & 0 & 0 \\ 0 & 0 & a \end{pmatrix} + \begin{pmatrix} 0 & 0 & 0 \\ 0 & b & 0 \\ 0 & 0 & 0 \end{pmatrix} + \begin{pmatrix} 0 & 0 & c \\ 0 & 0 & 0 \\ c & 0 & 0 \end{pmatrix}$$

$$= a \underbrace{\begin{pmatrix} 1 & 0 & 0 \\ 0 & 0 & 0 \\ 0 & 0 & 1 \end{pmatrix}}_{M_1} + b \underbrace{\begin{pmatrix} 0 & 0 & 0 \\ 0 & 1 & 0 \\ 0 & 0 & 0 \end{pmatrix}}_{M_2} + c \underbrace{\begin{pmatrix} 0 & 0 & 1 \\ 0 & 0 & 0 \\ 1 & 0 & 0 \end{pmatrix}}_{M_3},$$

which shows that E is spanned by $\{M_1, M_2, M_3\}$. Let us then show that the latter set is free. Let $\alpha, \beta, \gamma \in \mathbb{R}$ such that $\alpha M_1 + \beta M_2 + \gamma M_3 = 0_{M_3(\mathbb{R})}$. Then, we easily obtain that $\alpha = \beta = \gamma = 0$, which proves that $\{M_1, M_2, M_3\}$ is a basis of E. Thus, $\dim E = 3$.

Exercise 2.3.35. Let

$$E = \{(x,x) : x \in \mathbb{R}\}, \ F = \{(y,0) : y \in \mathbb{R}\}, \text{ and}$$
$$G = \{(0,z) : z \in \mathbb{R}\}.$$

(1) Show that $E \oplus F = \mathbb{R}^2$.
(2) Show that $E \oplus G = \mathbb{R}^2$.

Solution 2.3.35.

(1) Let $(x,y) \in \mathbb{R}^2$. We can express (x,y) uniquely as

$$(x,y) = (y,y) + (x-y,0).$$

Therefore, $E \oplus F = \mathbb{R}^2$. If readers are not fully convinced of the uniqueness of this decomposition (and they might not be), we can start from $E + F = \mathbb{R}^2$, which follows directly from the way we split the vector (x,y) above. To ensure the direct sum, we must also verify that $E \cap F = \{0\}$, which we can establish by showing that any vector in both E and F must be the zero vector.

Let us show that $E \cap F = \{0_{\mathbb{R}^2}\} = \{(0,0)\}$, which will confirm that $E \oplus F = \mathbb{R}^2$. Since we always have $\{(0,0)\} \subset E \cap F$, it remains to show the other inclusion. If $(x,y) \in E \cap F$, then $(x,y) \in E$ and $(x,y) \in F$. This implies $x = y$ and $y = 0$, leading to $x = y = 0$. Hence, $\{(0,0)\} \supset E \cap F$. Thus, $E \oplus F = \mathbb{R}^2$.

Here is another method that can be quite practical in some situations. We have

$$E = \{(x,x) : x \in \mathbb{R}\} = \{x(1,1) : x \in \mathbb{R}\}.$$

So, E is a vector space spanned by the *nonzero* vector $(1,1)$, thereby $\dim E = 1$. We also have

$$F = \{(y,0) : y \in \mathbb{R}\} = F = \{y(1,0) : y \in \mathbb{R}\},$$

which says that F is a vector space spanned by the *nonzero* vector $(1,0)$. Hence, $\dim F = 1$. As $E \cap F = \{(0,0)\}$, we can see that

$$\dim E + \dim F = 1 + 1 = 2 = \dim \mathbb{R}^2$$

and we conclude that $E \oplus F = \mathbb{R}^2$.

(2) We can apply either the first or second method found in the previous answer (with minor modifications) to show that $E \oplus G = \mathbb{R}^2$ (you can do it!).

Exercise 2.3.36. Let E and F be two subspaces of $\mathbb{R}^3$ given by:
$$E = \{(x, y, 0) : x, y \in \mathbb{R}\} \text{ and } F = \{(0, y, z) : z \in \mathbb{R}\}.$$
Do we have $E \oplus F = \mathbb{R}^3$?

Solution 2.3.36. Let $(x, y, z) \in \mathbb{R}^3$. We can write:
$$(x, y, z) = (x, y/2, 0) + (0, y/2, z),$$
which implies that $E + F = \mathbb{R}^3$. It's important to note that the way we have split the vector (x, y, z) is not unique. For example, the vector $(1, 2, 3)$ in $\mathbb{R}^3$ can be expressed in different forms, such as:
$$(1, 2, 3) = (1, 0, 0) + (0, 2, 3)$$
and
$$(1, 2, 3) = (1, 3, 0) + (0, -1, 3).$$
Hence, we "only" have: $E + F = \mathbb{R}^3$.

Exercise 2.3.37. Let E and F be two subspaces of $\mathbb{R}^3$ defined by:
$$E = \{(x, y, 0) : x, y \in \mathbb{R}\} \text{ and } F = \{(0, 0, z) : z \in \mathbb{R}\}.$$
Show that $E \oplus F = \mathbb{R}^3$.

Solution 2.3.37. Let $(x, y, z) \in \mathbb{R}^3$. We can express
$$(x, y, z) = (x, y, 0) + (0, 0, z)$$
in a unique way. Thus, $E \oplus F = \mathbb{R}^3$.

Exercise 2.3.38. Let E and F be two subspaces of $\mathbb{R}^3$ defined by:
$$E = \{(x, y, z) : x = y = z\} \text{ and } F = \{(0, y, z) : z \in \mathbb{R}\}.$$
Prove that $E \oplus F = \mathbb{R}^3$.

Solution 2.3.38. Let $(x, y, z) \in \mathbb{R}^3$. Then
$$(x, y, z) = \underbrace{(x, x, x)}_{\in E} + \underbrace{(0, y - x, z - x)}_{\in F},$$
meaning that $E + F = \mathbb{R}^3$. Since the decomposition above is clearly unique, the sum is a direct sum, i.e., $E \oplus F = \mathbb{R}^3$.

Exercise 2.3.39. Let E be a vector space and let F, G, and H be subspaces of E. Is it true that
$$(F + G + H = E \text{ and } F \cap G = \{0_E\}, F \cap H = \{0_E\}, G \cap H = \{0_E\})$$
$$\Longleftrightarrow F \oplus G \oplus H = E?$$

Solution 2.3.39. The answer is negative. Let $E = \mathbb{R}^3$, then consider
$$F = \{(x, y, 0) : \; x, y \in \mathbb{R}\}, \; G = \{(0, 0, z) : \; z \in \mathbb{R}\},$$
and
$$H = \{(0, y, y) : \; y \in \mathbb{R}\}.$$
Then F, G, and H are three subspaces of $\mathbb{R}^3$, which obey
$$F \cap G = \{(0, 0, 0)\}, F \cap H = \{(0, 0, 0)\}, \text{ and } G \cap H = \{(0, 0, 0)\}.$$
We also have $\mathbb{R}^3 = F + G + H$ because for all $(x, y, z) \in \mathbb{R}^3$
$$(x, y, z) = (x, y/2, 0) + (0, 0, z - y/2) + (0, y/2, y/2),$$
yet the sum is not direct since, for example,
$$(1, 1, 1) = \underbrace{(1, 2, 0)}_{\in F} + \underbrace{(0, 0, 2)}_{\in G} + \underbrace{(0, -1, -1)}_{\in H}$$
and
$$(1, 1, 1) = \underbrace{(1, 3, 0)}_{\in F} + \underbrace{(0, 0, 3)}_{\in G} + \underbrace{(0, -2, -2)}_{\in H}.$$

Exercise 2.3.40. Let E be the vector space of convergent real sequences in $\mathbb{R}$. Let F be the subspace of E consisting of constant sequences, and let G be the subspace of E of convergent sequences to 0. Show that
$$E = F \oplus G.$$

Solution 2.3.40. Let us prove that $F \cap G = \{0\}$ and $F + G = E$. We always have $\{0\} \subset F \cap G$, so we need to show that $F \cap G \subset \{0\}$. Let (u_n) be a sequence in both F and G. Since (u_n) is constant and converges to 0, we have $u_n = 0$ for all n, which means that $F \cap G \subset \{0\}$.

To prove that $F + G = E$, we start with a sequence (u_n) in E that converges to a limit ℓ. We define a new sequence (v_n) as $v_n = u_n - \ell$, which converges to 0 and thus belongs to G. We then consider the constant sequence (w_n) with values equal to ℓ. By expressing u_n as $u_n = w_n + v_n$, it is seen that E can be expressed as the sum of F and G, leading to the conclusion $E = F \oplus G$.

Exercise 2.3.41. Let E be the vector space of all functions defined from $\mathbb{R}$ to $\mathbb{R}$, and let
$$F = \{f : \mathbb{R} \to \mathbb{R} : \; f(-x) = f(x), \; \forall x \in \mathbb{R}\}$$
$$\text{and } G = \{f : \mathbb{R} \to \mathbb{R} : \; f(-x) = -f(x), \; \forall x \in \mathbb{R}\}.$$
(1) Prove that F and G are two subspaces of E.
(2) Show that
$$F \oplus G = E.$$

Solution 2.3.41.

(1) The set F represents all even functions, and G represents all odd functions. Readers should be able to quickly verify that F and G are two subspaces of E.

(2) Let f be a function in E. We need to demonstrate that f can be expressed (uniquely) as the sum of two functions, one even and the other odd. For all x in $\mathbb{R}$, we may write

$$f(x) = \underbrace{\frac{f(-x) + f(x)}{2}}_{g(x)} + \underbrace{\frac{f(-x) - f(x)}{2}}_{h(x)},$$

where g is an even function because

$$\forall x \in \mathbb{R}: \ g(-x) = \frac{f(x) + f(-x)}{2} = g(x),$$

and h is odd since

$$\forall x \in \mathbb{R}: \ h(-x) = \frac{f(x) - f(-x)}{2} = -\frac{f(-x) - f(x)}{2} = -h(x).$$

Thus, we have shown that $E = F + G$. To become a direct sum, we need further to show that $F \cap G = \{0\}$. As usual, we need only show that $\{0\} \supset F \cap G$. To this end, let $f \in F \cap G$, that is, $f \in F$ and $f \in G$. Then, for all x in $\mathbb{R}$, $f(-x) = f(x) = -f(x)$, i.e., for all $x \in \mathbb{R}$, $f(x) = 0$, that is, $f = 0$. Therefore, $F \cap G \subset \{0\}$, which yields $F \oplus G = E$, as wished.

Exercise 2.3.42. On $\mathbb{R}_3[X]$, consider the polynomials:

$$P_1 = X(X-1), \ P_2 = (X-1)^3, \text{ and } P_3 = X^3.$$

(1) Prove that P_1, P_2, and P_3 are linearly independent.
(2) Explain why the set $\{P_1, P_2, P_3\}$ is not a basis of $\mathbb{R}_3[X]$ and complete this set to a basis of $\mathbb{R}_3[X]$.
(3) Find a (direct) complement of the subspace F in $\mathbb{R}_3[X]$, where F is spanned by the vectors P_1 and P_2.

Solution 2.3.42.

(1) Let $a, b, c \in \mathbb{R}$ be such that $aP_1 + bP_2 + cP_3 = 0$, which means that

$$aP_1(X) + bP_2(X) + cP_3(X) = 0$$

for all $X \in \mathbb{R}$, that is,

$$aX(X-1) + b(X-1)^3 + cX^3 = 0,$$

still for all $X \in \mathbb{R}$. Assigning the values $-1, 0, 1$ to X results in the following system of three unknowns and three equations:

$$\begin{cases} 2a - 8b - c = 0, \\ b = 0, \\ c = 0. \end{cases}$$

Thus, $a = b = c = 0$, i.e., P_1, P_2, and P_3 are linearly independent in $\mathbb{R}^3$.

(2) The dimension of $\mathbb{R}_3[X]$ is four, so a set with only three elements, such as $\{P_1, P_2, P_3\}$, cannot be a basis of $\mathbb{R}_3[X]$. However, since the elements of $\{P_1, P_2, P_3\}$ are linearly independent, we can use the incomplete basis theorem to add a fourth vector and form a basis of $\mathbb{R}_3[X]$. Let us choose $P_0 = X$ as the fourth vector. We need to check that the elements of $\{P_1, P_2, P_3, P_0\}$ are also linearly independent in $\mathbb{R}_3[X]$ by showing that for any $a, b, c, d \in \mathbb{R}$, satisfying the equation $aP_1 + bP_2 + cP_3 + dP_0 = 0$, the values of $a, b, c,$ and d must be zero.

We have

$$aP_1(X) + bP_2(X) + cP_3(X) + dP_0(X) = 0,$$

for all $X \in \mathbb{R}$. In other words, $aX(X-1) + b(X-1)^3 + cX^3 + dX = 0$, still for all $X \in \mathbb{R}$. We can then create a system of four unknowns and equations by assigning four different values to X, for example, $-1, 0, 1, 2$. We then obtain the following system

$$\begin{cases} 2a - 8b - c - d = 0, \\ b = 0, \\ c + d = 0, \\ 2a + b + 8c + 2d = 0, \end{cases}$$

which leads to $a = b = c = d = 0$, indicating that $\{P_1, P_2, P_3, P_0\}$ is a basis of $\mathbb{R}_3[X]$.

(3) Let $E = \text{span}\{P_1, P_2\}$. Let $F = \text{span}\{P_3, P_0\}$ be its (direct) complement. We easily check that the families $\{P_1, P_2\}$ and $\{P_3, P_0\}$ are free. Their union gives $\{P_1, P_2, P_3, P_0\}$, which forms a basis of $\mathbb{R}_3[X]$, so $E \oplus F = \mathbb{R}_3[X]$.

Exercise 2.3.43. Let

$$E = \{A \in M_n(\mathbb{R}) : \ A^t = A\} \text{ and } F = \{A \in M_n(\mathbb{R}) : \ A^t = -A\}.$$

(1) Prove that E and F are two subspaces of $M_n(\mathbb{R})$.
(2) Find $\dim E$ and $\dim F$.
(3) Prove that
$$E \oplus F = M_n(\mathbb{R}).$$

Solution 2.3.43.

(1) We demonstrate that E is a subspace, leaving the proof that F is a vector space to readers. Let $\alpha, \beta \in \mathbb{R}$ and $A, B \in E$ (so $A^t = A$ and $B^t = B$). We want to show that $\alpha A + \beta B \in E$, i.e., $(\alpha A + \beta B)^t$. Using the properties of matrix transposition, we have:

$$(\alpha A + \beta B)^t = (\alpha A)^t + (\beta B)^t = \alpha A^t + \beta B^t = \alpha A + \beta B.$$

Therefore, $\alpha A + \beta B \in E$.

(2) To find a basis for E, we can follow these steps:
   - For the diagonal elements, we can independently choose each diagonal element. This means there are n independent choices for the diagonal elements. The corresponding basis matrices will have a 1 in one diagonal position and 0 elsewhere.
   - For the off-diagonal elements, due to the condition $A^t = A$, specifying the element a_{ij} (where $i < j$) automatically determines a_{ji}. Thus, there are $n(n-1)/2$ such pairs, each contributing one independent entry.

Thus, the total number of such matrices, which corresponds to the dimension of the space (once we show they are linearly independent, which is easy), is:

$$n + \frac{n(n-1)}{2} = \frac{n(n+1)}{2} = \dim E.$$

Similarly, we can prove that $\dim F = n(n-1)/2$. In a matrix A that obeys $A^t = -A$, all diagonal elements are zero, so they do not affect the dimension. The off-diagonal elements satisfy $a_{ij} = -a_{ji}$. For each pair of indices (i, j) with $i < j$, we can freely choose the value of a_{ij} (and then a_{ji} is automatically determined as $-a_{ij}$). Therefore, there are $n(n-1)/2$ such pairs.

(3) First, we show that: $E + F = M_n(\mathbb{R})$. Let A be an arbitrary matrix of $M_n(\mathbb{R})$. We can write:

$$A = \underbrace{\frac{A + A^t}{2}}_{B} + \underbrace{\frac{A - A^t}{2}}_{C}.$$

Let us check that $B \in E$ and $C \in F$. We have:

$$\left(\frac{A + A^t}{2} \right)^t = \frac{A^t + (A^t)^t}{2} = \frac{A^t + A}{2},$$

so $B \in E$. Likewise,

$$\left(\frac{A - A^t}{2}\right)^t = \frac{A^t - (A^t)^t}{2} = \frac{A^t - A}{2} = -\frac{A - A^t}{2},$$

meaning that $C \in F$. Thus, $E + F = M_n(\mathbb{R})$.

We now show that this sum is, in fact, a direct sum. Let $A \in E \cap F$, and we check that $A = 0_{M_n(\mathbb{R})}$. Since $A \in E \cap F$, we get $A = A^t$ and $A = -A^t$. Then: $A = -A$, i.e., $2A = 0_{M_n(\mathbb{R})}$ and so $A = 0_{M_n(\mathbb{R})}$. Therefore, we have shown that:

$$E \oplus F = M_n(\mathbb{R}).$$

Exercise 2.3.44. Determine the rank of each family of the given vectors in each of the following cases:
 (1) $u = (1, 0, 1)$, $v = (0, 1, 0)$, and $w = (0, 1, 1)$ in $\mathbb{R}^3$.
 (2) $u = (1, 1)$, $v = (2, 2)$, $w = (3, 3)$, and $z = (2, 0)$ in $\mathbb{R}^2$.
 (3) $u = (0, 0, 0)$, $v = (1, 2, 3)$, $w = (1, 0, 1)$ in $\mathbb{R}^3$.
 (4) $u = (1, 1, 1)$, $v = (2, 2, 2)$, $w = (3, 3, 3)$ in $\mathbb{R}^3$.
 (5) $u = (1, 1, 0, 0)$, $v = (0, 0, 2, 2)$, $w = (3, 3, 0, 0)$, and $z = (0, 0, 4, 4)$ in $\mathbb{R}^4$?

Solution 2.3.44. First, recall that the rank of a set of vectors is the maximum number of linearly independent vectors within that set.

 (1) It is easy to show that the three vectors are linearly *independent*. So, rank$\{u, v, w\} = 3$.
 (2) The dimension of $\mathbb{R}^2$ is 2, which means the rank of any set of vectors is less than or equal to 2, so it's either 1 or 2. In this case, the vectors u, v, and w are pairwise linearly dependent, while w and z are linearly independent. Therefore, the rank of the set $\{u, v, w, z\}$ is 2.
 (3) For a similar reason as before, the rank of this set of vectors is less than or equal to 3 as $\dim \mathbb{R}^3 = 3$. Since the three given vectors are linearly dependent due to u being the zero vector, the rank is less than or equal to 2. As v and w are not proportional, they are linearly independent. Therefore, the rank of the set $\{u, v, w\}$ is 2.
 (4) The rank is less than or equal to 3 because $\dim \mathbb{R}^3 = 3$. The three vectors are clearly linearly dependent, so their rank is less than or equal to 2.

    We also notice that (u and v), (v and w), and (u and w) are linearly dependent. Therefore, the rank is less than or equal to 1. The rank here is 1 since at least one vector is non-null, i.e., rank$\{u, v, w\} = 1$.

(5) The rank is necessarily less than or equal to four because $\dim \mathbb{R}^4 = 4$. The four vectors are linearly dependent, for example, $v = 2u$. This means that the rank is less than or equal to 3. We also notice that all combinations of three vectors are linearly dependent, so the rank is less than or equal to 2. We observe that u and w, as well as v and z, are linearly dependent. However, u and v are linearly independent, so we have found two linearly independent vectors. Therefore, $\text{rank}\{u, v, w, z\} = 2$.

Exercise 2.3.45. Using the method of Gaussian elimination, find the rank of the following matrices:

$$A = \begin{pmatrix} 1 & 2 & 1 & 2 \\ -2 & -3 & 0 & -5 \\ 4 & 9 & 6 & 7 \\ 1 & -1 & -5 & 5 \end{pmatrix}, \quad B = \begin{pmatrix} 1 & a & a^2 & a^3 \\ a & a^2 & a^3 & 1 \\ a^2 & a^3 & 1 & a \\ a^3 & 1 & a & a^2 \end{pmatrix},$$

where $a \in \mathbb{R}$.

Solution 2.3.45. We apply Gaussian elimination to the matrix until it reaches an echelon form. The row transformations $R_2 + 2R_1 \to R_2$, $R_3 - 4R_1 \to R_3$, and $R_4 - R_1 \to R_4$ yield the following matrix:

$$\begin{pmatrix} 1 & 2 & 1 & 2 \\ -2 & -3 & 0 & -5 \\ 4 & 9 & 6 & 7 \\ 1 & -1 & -5 & 5 \end{pmatrix} \rightarrow \begin{pmatrix} 1 & 2 & 1 & 2 \\ 0 & 1 & 2 & -1 \\ 0 & 1 & 2 & -1 \\ 0 & -3 & -6 & 3 \end{pmatrix}.$$

At this point, the matrix is not yet in echelon form because the second and third rows are identical, and the fourth row does not have a leading 1. To continue, we perform the transformations $R_3 - R_2 \to R_3$ and $R_4 + 3R_2 \to R_4$, resulting in:

$$\begin{pmatrix} 1 & 2 & 1 & 2 \\ 0 & 1 & 2 & -1 \\ 0 & 0 & 0 & 0 \\ 0 & 0 & 0 & 0 \end{pmatrix}.$$

Now, the matrix is in echelon form, and we conclude that the rank of the matrix is 2, as there are two nonzero rows.

In a similar manner, performing Gaussian elimination on B, we arrive at the following reduced form:

$$\begin{pmatrix} 1 & a & a^2 & a^3 \\ 0 & 1-a^4 & a(1-a^4) & a^2(1-a^4) \\ 0 & 0 & 1-a^4 & a(1-a^4) \\ 0 & 0 & 0 & 1-a^4 \end{pmatrix}.$$

Now, we analyze the rank of the matrix B:

- If $a^4 = 1$, i.e., $a = 1$ or $a = -1$, then the rank of B is 1, as the rows become linearly dependent.
- If $a^4 \neq 1$, i.e., $a \neq 1$ and $a \neq -1$, then the rank of B is 4, since the rows are linearly independent.

2.4. Exercises without Solutions

Exercise 2.4.1. Let E be the set of functions f defined from $[0,1]$ to $\mathbb{R}$ such that $f(1/2) = 2f(1)$. Is E a vector space over $\mathbb{R}$?

Exercise 2.4.2. Let E be the $\mathbb{R}$-vector space of continuous functions from $[0,\infty)$ to $\mathbb{R}$. Is the set $F = \{f \in E : \lim_{x\to\infty} f(x) = 0\}$ a subspace of E?

Exercise 2.4.3. Let I be any set, whether finite or infinite (countable or uncountable). Let $\{F_i : i \in I\}$ be a family of subspaces of a vector space E. Show that the union $\bigcup_{i\in I} F_i$ is a subspace of E.

Exercise 2.4.4. Among the following sets, determine which are subspaces of $\mathbb{R}^3$ and which are not. Provide reasons for your answers.

(1) $E = \{(x,y,z) \in \mathbb{R}^3 : x - y = 0\}$.
(2) $E = \{(x,y,z) \in \mathbb{R}^3 : \cos x = 1\}$.
(3) $E = \{(x,y,z) \in \mathbb{R}^3 : |x| = |y| = |z|\}$.

Exercise 2.4.5. Describe all subspaces of $\mathbb{R}^3$.

Exercise 2.4.6. Is $\{(x,y) \in \mathbb{Z}_5^2 : x + y = 0\}$ a subspace of $\mathbb{Z}_5^2$?

Exercise 2.4.7. Is $F = \{(z, z') \in \mathbb{C}^2 : z^2 + z'^2 = 0\}$ a subspace of $\mathbb{C}^2$?

Exercise 2.4.8. Determine whether the following families are linearly independent in their respective vector spaces:

(1) $\{(1,0,2,3), (3,2,0,1), (4,5,0,6), (6,0,5,4)\}$ in $\mathbb{R}^4$.
(2) $\{(1,2,0), (3,4,8), (-1,-3,0)\}$ in $\mathbb{R}^3$.
(3) $\{(1,-i), (2+i, 1), (4,0)\}$ in $\mathbb{C}^2$, where i denotes the imaginary unit.

Exercise 2.4.9. Let u, v, and w be three linearly independent vectors. Show that $u + v$, $v + w$, and $w + u$ are also linearly independent.

Exercise 2.4.10. Find a basis of the following subspace:
$$E = \{(x, y, z) \in \mathbb{R}^3 : 2x - y + 4z = 0\}.$$

Exercise 2.4.11. Determine the dimension of the subspace E in each of the following cases:
(1) $E = \{(x, y, z, t) \in \mathbb{R}^4 : t = 0\}$.
(2) $E = \{(x, y, z, t) \in \mathbb{R}^4 : t = x + y$ and $z = x - y\}$.
(3) $E = \{(x, y, z, t) \in \mathbb{R}^4 : x = y = z = t\}$?

Exercise 2.4.12. Consider the vectors
$$u = (-1, 2, 3) \text{ and } v = (1, -2, -2).$$
Extend the set $\{u, v\}$ to a basis of $\mathbb{R}^3$.

Exercise 2.4.13. Let $\mathbb{C}$ be a vector space over $\mathbb{R}$ and let $z = a + i b$. Find the conditions under which the set $\{z, \bar{z}\}$ forms a basis of $\mathbb{C}$. What are the coordinates of the complex number $x + i y$ in this new basis?

Exercise 2.4.14.
(1) Prove that the vectors
$$u = (1, 2 i, - i), \ v = (2, 1 + i, 1), \text{ and } w = (-1, 1, - i)$$
form a basis of the complex vector space $\mathbb{C}^3$.
(2) Find the coordinates of the vector $(1, 2, 0)$ with respect to this basis.

Exercise 2.4.15. Determine the coordinates of the vector $(2, -1, 3)$ with respect to the basis formed by the following vectors:
$$u = (1, 0, 0), \ v = (2, 2, 0), \text{ and } w = (3, 3, 3).$$

Exercise 2.4.16. Let $\{P_1, P_2, \ldots, P_n\}$ be a set of complex polynomials such that:
$$\deg P_1 < \deg P_2 < \cdots < \deg P_n.$$
Show that the elements of $\{P_1, P_2, \ldots, P_n\}$ are linearly independent.

Exercise 2.4.17. Let $E = C([0, 1])$ be the vector space of real continuous functions defined on $[0, 1]$. Let F be the set of constant functions on $[0, 1]$. Let G be the set of functions $g \in E$ satisfying:
$$\int_0^1 g(x) dx = 0.$$
(1) Show that F and G are two subspaces of E.
(2) Show that $F \oplus G = E$.

Exercise 2.4.18.

(1) Prove that $\{1, \pi, \pi^2, \ldots, \pi^n\}$ is a free set over $\mathbb{Q}$.

(2) Prove that $\mathbb{R}$, as a vector space over $\mathbb{Q}$, is infinite-dimensional.

Exercise 2.4.19. What is the rank of the following set of vectors $u = (1, 2, 3)$, $v = (3, 2, 1)$, $w = (2, 1, -3)$, and $z = (-4, -2, 6)$?

Exercise 2.4.20. Using the method of Gaussian elimination, find the rank of the matrix:

$$A = \begin{pmatrix} 1 & 1 & 1 \\ -1 & 2 & 2 \\ -1 & 5 & 5 \end{pmatrix}.$$

Exercise 2.4.21. What is the rank of the square matrix of order $n \geq 2$:

$$A = \begin{pmatrix} a_1 & a_2 & \cdots & a_n \\ a_2 & 0 & \cdots & 0 \\ \vdots & & \ddots & \vdots \\ a_n & 0 & \cdots & 0 \end{pmatrix},$$

where $a_1, a_2, \ldots, a_n \in \mathbb{R}$ and $a_2 \neq 0$?

CHAPTER 3

Linear Maps on Vector Spaces

3.1. Course Summary

DEFINITION 3.1.1. Let $f : E \to F$ be a mapping of vector spaces. We say that f is a linear map if:

- For all $x, y \in E$: $f(x + y) = f(x) + f(y)$.
- For any scalar α and any $x \in E$: $f(\alpha x) = \alpha f(x)$.

The set of linear maps from E to F is usually denoted as $L(E, F)$ and as $L(E)$ in case $E = F$.

Remark. The map f is not linear if any (or both!) of the two above conditions is not satisfied. More precisely, f is not linear if we can find two specific x and y such that $f(x + y) \neq f(x) + f(y)$, or if we can find a particular scalar α and a certain x such that $f(\alpha x) \neq \alpha f(x)$.

Remark. A linear map $f : E \to E$ is usually referred to as an endomorphism.

EXAMPLES 3.1.1.

(1) Let E be a vector space. The identity map $I : E \to E$ defined by $I(x) = x$ is linear.
(2) The map $f : \mathbb{R} \to \mathbb{R}$ defined by $f(x) = |x|$ is not linear.
(3) The map $f : \mathbb{R}^n \to \mathbb{R}^m$ defined by $f(x) = Ax$, where A is an $m \times n$ matrix, is linear.

A linear map from a vector space to its underlying field, which possesses interesting properties and applications here and elsewhere, has a particular appellation.

DEFINITION 3.1.2. Let E be vector space over a field $\mathbb{F}$. A linear functional (or linear form) f is a linear map from E to $\mathbb{F}$.

Next, we introduce two important concepts.

DEFINITION 3.1.3. Let $f \in L(E, F)$, where E and F are two vector spaces.

(1) The **kernel** of f (or the **null-space** of f) is defined by

$$\ker f = \{x \in E : f(x) = 0_F\}.$$

(2) The **range** of f (or **image** of f) is defined as

$$\operatorname{ran} f = f(E) = \{f(x) : \ x \in E\}.$$

PROPOSITION 3.1.1. *Let $f \in L(E, F)$, where E and F are two vector spaces. Then $\ker f$ is a linear subspace of E, and $\operatorname{ran} f$ is a linear subspace of F.*
Moreover, f is injective if and only if $\ker f = \{0_E\}$.

Thus, we may introduce the following notations:

DEFINITION 3.1.4. Let $f \in L(E, F)$, where E and F are two vector spaces.

(1) The dimension of $\ker f$ is called the **nullity** of f, that is, the number $n(A) = \dim(\ker f)$.
(2) The dimension of $\operatorname{ran} f$ is called the **rank** of f, that is, the number $\dim(\operatorname{ran} f)$.

DEFINITION 3.1.5. Let E and F be two vector spaces and $f : E \to F$ be a linear transformation. Call f an isomorphism if it is bijective, i.e., f is injective and surjective.

PROPOSITION 3.1.2. *Let E and F be two vector spaces and let $f : E \to F$ be an isomorphism. Then $f^{-1} : F \to E$ is also an isomorphism, i.e., it is linear and bijective.*

The following result is deeper.

THEOREM 3.1.3. *Let E and F be two finite-dimensional vector spaces over the same field and $f : E \to F$ be a linear transformation. Then, the following properties are equivalent:*

(1) *f is an isomorphism.*
(2) *$\dim E = \dim F$ and $\ker f = \{0_E\}$.*

3.1.1. Rank-nullity theorem.

One of the most fundamental theorems in linear algebra, more commonly known as the rank-nullity theorem, is recalled next:

THEOREM 3.1.4. *(Cf. Exercise 3.3.34) Let $f : E \to F$ be a linear transformation between vector spaces, where $\dim E$ is finite. Then, $\dim(\ker f)$ and $\dim f(E)$ are both finite-dimensional, and we have:*

$$\operatorname{rank}(f) + \operatorname{nullity}(f) = \dim E.$$

3.1.2. Associated matrices with linear maps.

Let E and F be two vector spaces of dimensions n and m respectively. Let $B = \{e_1, \ldots, e_n\}$ be a basis for E and $B' = \{e'_1, \ldots, e'_m\}$ be a basis for F. Let $f : E \to F$ be a linear transformation. It is known that for each j, we may write $f(e_j) = \sum_{i=1}^{m} a_{ij} e'_i$, where $a_{ij} \in \mathbb{F}$. These scalars are in fact the entries of the associated matrix, noted A. In other words,

$$A = \begin{pmatrix} a_{11} & a_{12} & \cdots & a_{1n} \\ a_{21} & a_{22} & \cdots & a_{2n} \\ \vdots & \vdots & \ddots & \vdots \\ a_{m1} & a_{m2} & \cdots & a_{mn} \end{pmatrix},$$

which is of size $m \times n$.

EXAMPLE 3.1.1. Consider the linear map $f : \mathbb{R}^2 \to \mathbb{R}^2$ defined by:

$$f(x, y) = (x + y, 2x - y).$$

The standard basis for $\mathbb{R}^2$ is $\{(1,0), (0,1)\}$. Applying T to the basis vectors, we have:

$$f(1,0) = (1 + 0, 2 \cdot 1 - 0) = (1, 2),$$
$$f(0,1) = (0 + 1, 2 \cdot 0 - 1) = (1, -1).$$

The matrix representation of f with respect to the *standard* basis is:

$$A = \begin{pmatrix} 1 & 1 \\ 2 & -1 \end{pmatrix}.$$

3.1.3. Transition matrix.

Consider two bases $B = \{e_1, \ldots, e_n\}$ and $B' = \{e'_1, \ldots, e'_n\}$ of the same vector space E. A transition matrix (or change-of-basis matrix) from B to B' is the matrix P whose columns are the coordinates of the vectors in B expressed in the basis B'.

Remark. It is clear that transition matrices are square.

THEOREM 3.1.5. *A transition matrix from a basis to another is always invertible.*

COROLLARY 3.1.6. *If P is the transition matrix from B to B', then P^{-1} is the transition matrix from B' to B.*

Transition matrices are usually used to find the expression of a matrix with respect to a new basis. This is described in the following theorem:

THEOREM 3.1.7. *Let E be a vector space, and let $f : E \to E$ be a linear transformation. Suppose $B = \{e_1, e_2, \ldots, e_n\}$ and $B' = \{e'_1, e'_2, \ldots, e'_n\}$ are two bases of E. If the matrix of f in the basis B is A, and the transition matrix from B to B' is P, then the matrix A' of f in the basis B' is given by:*

$$A' = P^{-1} A P.$$

The previous theorem deals with linear maps from one vector space to itself, hence the associated matrices are square. Below, we present the more general case:

THEOREM 3.1.8. *Let E and F be vector spaces, and let $f : E \to F$ be a linear transformation. Suppose $B = \{e_1, e_2, \ldots, e_m\}$ and $C = \{f_1, f_2, \ldots, f_n\}$ are two bases of E and F, respectively. Consider two other bases of E and F, denoted $B' = \{e'_1, e'_2, \ldots, e'_m\}$ and $C' = \{f'_1, f'_2, \ldots, f'_n\}$, respectively.*

If the matrix of f in the bases B and C is A, and the transition matrix from B to B' is P, and the transition matrix from C to C' is Q, then the matrix A' of f in the bases B' and C' is given by:

$$A' = Q^{-1}AP.$$

3.2. True or False

Questions. Answer the following questions, providing justifications for your answers:

(1) Constant mappings are obviously linear.
(2) Let $f : \mathbb{R} \to \mathbb{R}$ be such that $f(x) = x^2$. Then f is linear.
(3) Let $f, g : \mathbb{R} \to \mathbb{R}$ be two linear maps. Then fg (defined for each $x \in \mathbb{R}$ by $(fg)(x) = f(x) \times g(x)$) is linear.
(4) Let E, F, and G be three vector spaces (over the same field $\mathbb{F}$). Let $f : E \to F$ and $g : F \to G$ be two linear maps. Then $g \circ f$ is linear.
(5) Let E and F be two vector spaces (over the same field $\mathbb{F}$). Let $f : E \to F$ be a map. Then

$$f \text{ is linear} \iff f(0_E) = 0_F.$$

(6) Let E and F be two vector spaces (over the same field $\mathbb{F}$). Let $f : E \to F$ be a map. Then

$$f \text{ is linear} \iff f(-x) = -f(x), \ \forall x \in E.$$

(7) Let E and F be two vector spaces (over the same field). Let $f : E \to F$ be a map. Then

$$f \text{ is injective} \iff \ker f = \{0_E\}.$$

(8) Let f be a map from E to F, where E and F are two vector spaces. Then $\ker f$ is a linear subspace of E, and $\operatorname{ran} f$ is a linear subspace of F.
(9) Let E and F be two vector spaces over the same field, and let $f : E \to F$ be a linear map. What can you say if $\ker f = E$? What about when $\operatorname{ran} f = \{0_F\}$?

(10) Let E and F be two vector spaces over the same field, both of finite dimension, and let $f : E \to F$ be a linear map. Then

$$\dim F = \dim \ker f + \dim \operatorname{ran} f.$$

(11) Let E and F be two vector spaces, where E does not necessarily have a finite dimension, and both E and F are defined over the same field. Let $f : E \to F$ be a linear map. Then,

$$\dim E = \dim \ker f + \dim \operatorname{ran} f.$$

(12) Let E and F be two vector spaces over the same field, both of finite dimension, and let $f : E \to F$ be a linear map. If $\dim E > \dim F$, then f cannot be injective.

(13) Let E and F be two vector spaces over the same field, both of finite dimension, and let $f : E \to F$ be a linear map. If $\dim E > \dim F$, then f is always surjective.

(14) Let $f, g : E \to E$ be two linear maps, i.e., two endomorphisms of E. Then,

$$g \circ f = 0 \iff \operatorname{ran} f \subset \ker g.$$

(15) Let f and g be two endomorphisms of E. Then

$$\operatorname{ran}(f + g) = \operatorname{ran} f + \operatorname{ran} g.$$

(16) Let $f, g : E \to E$ be two linear maps. Then

$$\operatorname{ran}(f \circ g) = \operatorname{ran}(g \circ f).$$

(17) Let E and F be two vector spaces and $f : E \to F$ be a linear transformation. Suppose $\{e_1, \ldots, e_n\}$ is a basis of E. Then $\{f(e_1), \ldots, f(e_n)\}$ is a basis of F.

(18) The matrix $A = \begin{pmatrix} 1 & 2 \\ 3 & i \end{pmatrix}$ can be defined on $\mathbb{C}^2$ as well as on $\mathbb{R}^2$.

(19) The matrix $A = \begin{pmatrix} 1 & 2 \\ 3 & 2 \end{pmatrix}$ may be defined on $\mathbb{R}^2$ as well as on $\mathbb{C}^2$.

Answers.

(1) **False!** Consider a mapping $f : E \to F$, where $f(x) = c$ for all $x \in E$, where c is a constant in F. For this mapping to be linear, it must at least satisfy $f(0) = 0$. However, if $c \neq 0_F$, the mapping would never be linear. Therefore, the only linear constant transformation is the null function.

(2) **False!** In this example, f is not linear because, for instance,

$$f(2 \times 3) = f(6) = 6^2 = 36 \neq 2 \times f(3) = 2 \times 3^2 = 18.$$

(3) **False!** For example, let us take $f(x) = g(x) = x$, both defined from $\mathbb{R}$ to $\mathbb{R}$, as a counterexample. Then f and g are both linear, but $(fg)(x) = x^2$ is not linear.

(4) **True.** Let $g \circ f : E \to G$ be defined by $(g \circ f)(x) = g(f(x))$. Let us show that $g \circ f$ is linear. Let $x, y \in E$ and $\alpha, \beta \in F$, we have:

$$(g \circ f)(\alpha x + \beta y) = g(f(\alpha x + \beta y)),$$

so

$$
\begin{aligned}
(g \circ f)(\alpha x + \beta y) &= g(\alpha f(x) + \beta f(y)) \text{ (because } f \text{ is linear)} \\
&= \alpha g(f(x)) + \beta g(f(x)) \text{ (because } g \text{ is linear)} \\
&= \alpha (g \circ f)(x) + \beta (g \circ f(y)).
\end{aligned}
$$

Remark. The composition of linear maps is often referred to as a "product" or "multiplication." This terminology arises because multiplying two matrices corresponds to composing the associated linear transformations. Specifically, if A is the matrix associated with a linear map $f : E \to F$, and B is the matrix associated with another linear map $g : F \to G$, then the matrix associated with the composition $g \circ f : E \to G$ is simply BA.

(5) **False!** Only the implication "$\Longrightarrow$" is true. This implication is commonly used to test whether a map is linear. By considering the contrapositive, we see that if $f(0_E) \neq 0_F$, then f is not linear. This observation is very convenient and is typically the first approach used when determining if a map is linear. However, no conclusion can be drawn if $f(0_E) = 0_F$.

To recap:

(a) If $f(0_E) \neq 0_F$, then f is not linear.

(b) The condition $f(0_E) = 0_F$ alone does not ensure that f is linear. So, additional conditions or methods must be employed to verify the linearity of f.

Now, the implication "$\Longleftarrow$" is not always true. For instance, consider the function $f(x) = x^2$ (defined from $\mathbb{R}$ to $\mathbb{R}$). Although f is not linear, we have $f(0) = 0$!

(6) **False!** Only "$\Longrightarrow$" is true. Thus, by contrapositive, if there is an $x \in E$ such that $f(-x) \neq -f(x)$, then f is *not* linear.

However, if $f(-x) = -f(x)$ for all $x \in E$, we cannot make any conclusions. For example, if we take $f(x) = x^3$ (defined from $\mathbb{R}$ to $\mathbb{R}$), then f is not linear even though for all $x \in E$: $f(-x) = (-x)^3 = -x^3 = -f(x)$.

To summarize:

(a) If there exists an $x \in E$ such that $f(-x) \neq -f(x)$, then f is not linear.

(b) If for all $x \in E$: $f(-x) = -f(x)$, then we cannot determine anything about the linearity of f, and we need to explore other ways to see whether f is linear.

(7) **False!** Consider the function $f : \mathbb{R} \to \mathbb{R}$ defined by $f(x) = x^2$. We have

$$\ker f = \{x \in \mathbb{R} : f(x) = 0\} = \{x \in \mathbb{R} : x^2 = 0\} = \{0\}.$$

However, f is not injective as $f(-2) = f(2) = 4$ while $2 \neq -2$.

(8) **False!** Indeed, a crucial ingredient—linearity—is absent. For example, consider the *nonlinear* $f : \mathbb{R} \to \mathbb{R}$ defined by $f(x) = (x - 1)^2$. It is then clear that $\ker f = \{1\}$, which is not a linear subspace of $\mathbb{R}$, and $\operatorname{ran} f = [0, \infty)$, which is not a linear subspace of $\mathbb{R}$ either!

(9) If $\ker f = E$, this implies that $f(x) = 0$ for all $x \in E$. In other words, f is the zero function.

Now, let us assume that a linear map f satisfies $\operatorname{ran} f = \{0_F\}$. This means that $f(x) = 0_F$ for all $x \in E$, and thus f is the zero function in this case as well.

(10) **False!** Let us supply a counterexample. Let $f : \mathbb{R}^2 \to \mathbb{R}^3$ be defined by $f(x, y) = (x, y, 0)$, which is clearly linear. Readers can readily check that $\ker f = \{(0, 0)\}$, giving $\dim \ker f = 0$. Also, $\dim \operatorname{ran} f = 2$. Thus,

$$\dim \mathbb{R}^3 = 3 \neq 2 = \dim \ker f + \dim \operatorname{ran} f,$$

as intended.

(11) **False!** The dimension of E must be *finite* in order to apply this result! Additionally, we will explore the significance of the "finite dimension" hypothesis in Exercises 3.3.28 and 3.3.29.

(12) **True.** Let us demonstrate that f is not injective. Suppose that it is injective. This implies that $\dim \ker f = 0$. According to the rank-nullity theorem, $\dim E = \dim \operatorname{ran} f$. However, since $\operatorname{ran} f$ is a subspace of F, we have $\dim \operatorname{ran} f \leq \dim F$. Consequently, $\dim E \leq \dim F$, which contradicts the hypothesis that $\dim E > \dim F$. Therefore, we conclude that f is not injective.

(13) **False!** For instance, let $f : \mathbb{R}^2 \to \mathbb{R}$ be defined by: $f(x) = 0$. Then f is not surjective, yet we see that $\dim \mathbb{R}^2 > \dim \mathbb{R}$.

(14) **True.** If $g \circ f = 0$, and y is in the range of f, then there exists an x in E such that $y = f(x)$. This implies that $g(y) = g(f(x)) = 0$, resulting in y being in the kernel of g.

Conversely, if the range of f is a subset of the kernel of g, then for any x in E, $f(x)$ is in the range of f and also in the kernel of g. Thus, $g \circ f = 0$.

(15) **False!** Only the inclusion $\operatorname{ran}(f + g) \subset \operatorname{ran} f + \operatorname{ran} g$ is true. Let us prove this. Let $y \in \operatorname{ran}(f + g)$, i.e., there exists $x \in E$ such that $y = (f + g)(x) = f(x) + g(x)$. Since $f(x) \in \operatorname{ran} f$ and $g(x) \in \operatorname{ran} g$, it follows that $y \in \operatorname{ran} f + \operatorname{ran} g$. Thus, the inclusion holds.

However, the reverse inclusion $\operatorname{ran}(f + g) \supset \operatorname{ran} f + \operatorname{ran} g$ is not necessarily true in general. For instance, consider a surjective endomorphism f of $\mathbb{R}$ and let $g = -f$. Then, $\operatorname{ran} f = \operatorname{ran} g = \mathbb{R}$, so $\operatorname{ran} f + \operatorname{ran} g = \mathbb{R}$, while $\operatorname{ran}(f+g) = \{0\}$.

(16) **False!** Consider two linear maps $f, g : \mathbb{R}^2 \to \mathbb{R}^2$ defined by

$$f(x, y) = (x, 0) \text{ and } g(x, y) = (0, x),$$

respectively. That f and g are linear may be easily verified by the readers. We have $(f \circ g)(x, y) = f(g(x, y)) = f(0, x) = (0, 0)$ and $(g \circ f)(x, y) = g(f(x, y)) = g(x, 0) = (0, x)$. From this, we can see that

$$\operatorname{ran}(f \circ g) = \{(0, 0)\} \neq \{0\} \times \mathbb{R} = \operatorname{ran}(g \circ f),$$

as wished.

(17) **False!** Let us provide a counterexample. Let $f : \mathbb{R} \to \mathbb{R}$ be defined by $f(x) = 0$, which is obviously linear. Then, $\{1\}$ is a basis of the domain $\mathbb{R}$, yet $f(\{1\}) = \{0\}$ is not a basis of the codomain $\mathbb{R}$. For more related properties, readers may consult Exercise 3.3.23.

(18) The matrix $A = \begin{pmatrix} 1 & 2 \\ 3 & i \end{pmatrix}$ can only be defined on $\mathbb{C}^2$. The matrix A has a complex entry, which implies that the underlying field over which the matrix operates should be the complex numbers $\mathbb{C}$. Therefore, A, which has one complex entry, cannot be consistently defined on $\mathbb{R}^2$ because it breaks the fundamental structure and operations defined over the field of real numbers.

(19) Both statements are true. The matrix A, which consists entirely of real entries, can be defined on $\mathbb{R}^2$ and $\mathbb{C}^2$ alike because the real numbers are a subset of the complex numbers, and

the operations defined in the real field extend naturally to the complex field. This ensures that the matrix operations remain consistent and valid in both vector spaces.

3.3. Exercises with Solutions

Exercise 3.3.1. Determine which of the following maps are linear from E to F (over $\mathbb{R}$):

(1) $E = F = \mathbb{R}^+$, $f(x) = 2x$, where $\mathbb{R}^+ = [0, \infty)$.

(2) $E = F = \mathbb{R}^2$; $f(x, y) = (x - y, 1 + x)$.

(3) $E = F = \mathbb{R}^2$; $f(x, y) = (x - 2y, x)$.

(4) $E = \mathbb{R}^2$, $F = \mathbb{R}^3$; $f(x, y) = (x - y, -1 + 2x, y)$.

(5) $E = F = \mathbb{R}^3$; $f(x, y, z) = (x - y + z, y - z, x^2)$.

(6) $E = F = \mathbb{R}^2$; $f(x, y) = \left(\frac{y}{x^2 + 1}, \frac{x}{2 + x^2 + y^2} \right)$.

(7) $E = \mathbb{R}$, $F = \mathbb{R}^2$; $f(x) = (|x|, 2x)$.

(8) $E = \mathbb{R}^3$, $F = \mathbb{R}^2$; $f(x, y, z) = (e^x, y \sin z)$.

(9) $E = \mathbb{R}^3$, $F = \mathbb{R}^2$; $f(x, y, z) = (ye^x, y \sin z)$.

Solution 3.3.1.

(1) The map f cannot be classified as linear because $\mathbb{R}^+$ is not a vector space.

(2) A linear map must map the zero vector in (the domain) $\mathbb{R}^2$ to the zero vector in (the codomain) $\mathbb{R}^2$. So, the map f is not linear since

$$f(0, 0) = (0 - 0, 0 + 1) = (0, 1) \neq (0, 0).$$

(3) This time, we do have $f(0, 0) = (0, 0)$, but we have yet to conclude. We need to expand the definition of linearity further. Let $(x, y), (x', y') \in \mathbb{R}^2$. Let $\alpha, \beta \in \mathbb{R}$. Then

$$
\begin{aligned}
f(\alpha(x, y) + \beta(x', y')) &= f((\alpha x, \alpha y) + (\beta x', \beta y')) \\
&= f(\alpha x + \beta x', \alpha y + \beta y') \\
&= (\alpha x + \beta x' - 2(\alpha y + \beta y'), \alpha x + \beta x') \\
&= (\alpha x + \beta x' - 2\alpha y - 2\beta y', \alpha x + \beta x') \\
&= (\alpha x - 2\alpha y, \alpha x) + (\beta x' - 2\beta y', \beta x') \\
&= \alpha(x - 2y, x) + \beta(x' - 2y', x') \\
&= \alpha f(x, y) + \beta f(x', y'),
\end{aligned}
$$

which indicates that f is linear.

(4) Since $f(0,0) = (0 - 0, -1 + 2 \times 0, 0) = (0, -1, 0) \neq (0, 0, 0)$, it follows that f is not linear.

(5) Even $f(0,0,0) = (0,0,0)$, we cannot draw any conclusions. Since, in general, $(x + x')^2 \neq x^2 + x'^2$, we can expect f not be linear. Let us corroborate this by exhibiting a counterexample. If f were linear, then we would have $f(U + V) = f(U) + f(V)$ for all (U, V). However, when $U = (2, 0, 0)$ and $V = (1, 0, 0)$, we get

$$f(U + V) = f((2,0,0) + (1,0,0)) = f((3,0,0)) = (3, 0, 9),$$

whereas

$$f(U) + f(V) = f((2,0,0)) + f((1,0,0))$$
$$= (2, 0, 4) + (1, 0, 1) = (3, 0, 5).$$

Thus,

$$f(U + V) \neq f(U) + f(V),$$

meaning that additivity fails, thereby f is not linear.

(6) The map f is not linear because it doesn't satisfy the additivity property. Indeed, while $f(1, -1) = (-1/2, 1/4)$, we can see that

$$f(1, -1) = f(1, 0) + f(0, -1) = (0, 1/3) + (-1, 0) = (-1, 1/3),$$

and so $f[(1, 0) + (0, -1)] \neq f(1, 0) + f(0, -1)$, which signifies that f is not linear.

(7) The map f is not linear because it does not satisfy the homogeneity property. Indeed, the fact that

$$f(-1 \times 2) = (|-2|, -2 \times 2) = (2, -4) \neq (-1) \times f(2) = (-2, -4)$$

clearly violates the homogeneity property, thus proving that the map f is not linear.

(8) The map f is not linear for $f(0, 0, 0) = (1, 0) \neq (0, 0)$.

(9) The map f is not linear. For instance,

$$f(-1, 1, 0) + f(1, 0, 0) = (e^{-1}, 0) + (0, 0) = (e^{-1}, 0),$$

whereas

$$f[(-1, 1, 0) + (1, 0, 0)] = f(0, 1, 0) = (1, 0) \neq (e^{-1}, 0).$$

Exercise 3.3.2. Determine the linearity of $T : E \to F$ in each of the following cases:

(1) $T(f)(x) = |f(x)|$, where $E = F = \mathcal{F}(\mathbb{R}, \mathbb{R})$ denote the vector space of functions defined from $\mathbb{R}$ to $\mathbb{R}$.

(2) $T(f)(x) = f(x^2)$, where $E = F = \mathcal{F}(\mathbb{R}, \mathbb{R})$.

(3) $T(f)(x) = e^x f(x)$, where $E = F = \mathcal{F}(\mathbb{R}, \mathbb{R})$.

(4) $T(f)(x) = f'(x)$, where $E = C^1(\mathbb{R}, \mathbb{R})$ and $F = C(\mathbb{R}, \mathbb{R})$.

(5) $T(f)(x) = \int_0^x f(t)dt$, where $E = F = C(\mathbb{R}, \mathbb{R})$.

Solution 3.3.2.

(1) The transformation T is nonlinear because it does not satisfy the homogeneity property. For instance, $T(-2f) \neq -2T(f)$, when, for example, $f(x) = 1$ for all x. To illustrate, we have

$$T(-2f(x)) = |-2f(x)| = 2 \neq -2 = -2|f(x)|.$$

(2) Before proceeding with the answer to this question, recall that we already know above that $x \mapsto x^2$, defined from $\mathbb{R}$ to $\mathbb{R}$, is not linear. However, in this case, we should not worry much about the term x^2, as we should focus on the variable f, and linearity needs to be checked with respect to it. That being said, the given transformation T is linear, and to see that, let $f, g \in E$ and $\alpha, \beta \in \mathbb{R}$. Then, for all $x \in \mathbb{R}$

$$\begin{aligned}
T(\alpha f + \beta g)(x) &= (\alpha f + \beta g)(x^2) \\
&= \alpha f(x^2) + \beta g(x^2) \\
&= \alpha T(f)(x) + \beta T(g)(x).
\end{aligned}$$

demonstrating the linearity of T.

(3) The map T is linear because for all $\alpha, \beta \in \mathbb{R}$ and all $f, g \in E$ (and each real x), we have:

$$\begin{aligned}
T(\alpha f + \beta g)(x) &= e^x(\alpha f + \beta g)(x) \\
&= \alpha e^x f(x) + \beta e^x g(x) \\
&= \alpha T f(x) + \beta T g(x).
\end{aligned}$$

(4) To show that T is linear, let $f, g \in E$ and $\alpha, \beta \in \mathbb{R}$. Then, for all x, we have, using the differentiability properties of functions under addition and scalar multiplication,

$$\begin{aligned}
T(\alpha f + \beta g)(x) &= (\alpha f + \beta g)'(x) = \alpha f'(x) + \beta g'(x) \\
&= \alpha T f(x) + \beta T g(x),
\end{aligned}$$

which establishes the linearity of T, as desired.

(5) The transformation T is linear thanks to the well-known properties of integrals. Indeed, let $\alpha, \beta \in \mathbb{R}$ and $f, g \in E$, then for all real x:

$$T(\alpha f + \beta g)(x) = \int_0^x (\alpha f(t) + \beta g(t))dt$$

$$= \alpha \int_0^x f(t)dt + \beta \int_0^x g(t)dt$$

$$= \alpha T f(x) + \beta T g(x),$$

which shows the linearity of T.

Exercise 3.3.3. Show that $f : E \to F$ is linear in each of the following cases:

(1) $E = \mathbb{R}_2[X]$, $F = \mathbb{R}$, and f is defined by $f(P) = P(2)$,
(2) $E = \mathbb{R}_2[X]$, $F = \mathbb{R}_3[X]$, and f is defined by

$$f(P) = (1 - X^2)P' + XP.$$

Solution 3.3.3.

(1) The mapping f is linear because

$$f(\alpha P + \beta Q) = (\alpha P + \beta Q)(2) = \alpha \underbrace{P(2)}_{f(P)} + \beta \underbrace{Q(2)}_{f(Q)},$$

which holds for all $P, Q \in E$ and for all $\alpha, \beta \in \mathbb{R}$.

(2) Let $P, Q \in E$ and let $\alpha, \beta \in \mathbb{R}$. Then

$$f(\alpha P + \beta Q) = (1 - X^2)(\alpha P + \beta Q)' + X(\alpha P + \beta Q)$$

$$= (1 - X^2)(\alpha P' + \beta Q') + X(\alpha P + \beta Q)$$

$$= (1 - X^2)(\alpha P') + X(\alpha P)$$

$$+ (1 - X^2)(\beta Q') + X(\beta Q)$$

$$= \alpha((1 - X^2)P' + XP) + \beta((1 - X^2)Q' + XQ)$$

$$= \alpha f(P) + \beta f(Q),$$

which indicates that f is linear.

Exercise 3.3.4. Which mappings from $\mathbb{R}$ to $\mathbb{R}$ are linear?

Solution 3.3.4. The only linear maps from $\mathbb{R}$ to $\mathbb{R}$ are of the form $f(x) = ax$, where $a \in \mathbb{R}$. Let us prove this statement.

(1) If $f(x) = ax$, where $a \in \mathbb{R}$, it is clear that f is linear.
(2) Let us prove that if the function f is linear from $\mathbb{R}$ to $\mathbb{R}$, then it must be of the form $f(x) = ax$, where $a \in \mathbb{R}$. Take any $x \in \mathbb{R}$. The linearity of f allows us to write $f(x) = f(x \times 1) = xf(1)$.

If we set $a = f(1)$, which is a real number, then for each x in $\mathbb{R}$, we have $f(x) = ax$.

Remark. Readers can also show that the general form of linear maps from $\mathbb{R}^2$ to $\mathbb{R}^2$ is

$$f(x, y) = (ax + by, cx + dy),$$

where $a, b, c, d \in \mathbb{R}$. What is the general form of linear maps from $\mathbb{R}^n$ to $\mathbb{R}^m$?

Exercise 3.3.5. What are the only linear transformations $f : \mathbb{R} \to \mathbb{R}$ that satisfy $f(x^2) = [f(x)]^2$ for all x in $\mathbb{R}$?

Solution 3.3.5. We know that the only linear mappings from $\mathbb{R}$ to $\mathbb{R}$ are given by $f(x) = ax$, where $a \in \mathbb{R}$. Thus, the hypothesis $f(x^2) = [f(x)]^2$ becomes nothing but $ax^2 = a^2x^2$ for all $x \in \mathbb{R}$. In particular, when $x = 1$, we end up with $a^2 = a$, which indicates that either $a = 0$ or $a = 1$. In consequence, the only sought linear maps are either $f(x) = 0$ or $f(x) = x$, both defined from $\mathbb{R}$ to $\mathbb{R}$.

Exercise 3.3.6. Let the following maps be defined from $\mathbb{C}$ to $\mathbb{C}$:

$$f(z) = \overline{z}, \ g(z) = \mathrm{Re}(z), \text{ and } h(z) = \mathrm{Im}(z).$$

(1) Is f $\mathbb{C}$-linear? $\mathbb{R}$-linear?
(2) Prove that g and h are $\mathbb{R}$-linear. Are they $\mathbb{C}$-linear?

Solution 3.3.6.

(1) Let $z, z' \in \mathbb{C}$ and let $\alpha, \beta \in \mathbb{C}$. We have:

$$f(\alpha z + \beta z') = \overline{\alpha z + \beta z'} = \overline{\alpha}\ \overline{z} + \overline{\beta}\ \overline{z'} \neq \alpha \overline{z} + \beta \overline{z'}$$

unless α and β are always real, which is not the case here. Thus, we observe that f is not $\mathbb{C}$-linear. To explicitly show this, consider a counterexample: For instance, let $\alpha = \mathrm{i}$ and $z = 1 + 2\,\mathrm{i}$. Then,

$$f(\mathrm{i}(1 + 2\,\mathrm{i})) = \overline{\mathrm{i}(1 + 2\,\mathrm{i})} = \overline{\mathrm{i} - 2} = -\mathrm{i} - 2 \neq \mathrm{i}(\overline{1 + 2\,\mathrm{i}}) = \mathrm{i} + 2.$$

Therefore, f is not $\mathbb{C}$-linear.

However, f is $\mathbb{R}$-linear. Indeed, let $z, z' \in \mathbb{C}$ and $\alpha, \beta \in \mathbb{R}$. Then

$$f(\alpha z + \beta z') = \overline{\alpha z + \beta z'} = \overline{\alpha}\ \overline{z} + \overline{\beta}\ \overline{z'} = \alpha \overline{z} + \beta \overline{z'}$$
$$= \alpha f(z) + \beta f(z'),$$

which shows that f is $\mathbb{R}$-linear.

(2) We will only consider the case of g (the case of h is left for interested readers). It is helpful to decompose z as $z = x + iy$. To show that g is $\mathbb{R}$-linear, let $z, z' \in \mathbb{C}$ and $\alpha, \beta \in \mathbb{R}$. We have:

$$g(\alpha z + \beta z') = g[\alpha(x + iy) + \beta(x' + iy')]$$
$$= g[\alpha x + \beta x' + i(\alpha y + \beta y')].$$

Then

$$g(\alpha z + \beta z') = \mathrm{Re}(\alpha z + \beta z') = \alpha x + \beta x' = \alpha \, \mathrm{Re}(z) + \beta \, \mathrm{Re}(z').$$

Now, g is not $\mathbb{C}$-linear because if it were, we would have, for example, $g(i \times i) = i \times g(i)$, which is not the case since

$$g(i \times i) = \mathrm{Re}(i \times i) = \mathrm{Re}(-1) = -1 \neq i \times \mathrm{Re}(i) = i \times 0 = 0.$$

Exercise 3.3.7. Suppose a linear map $f : \mathbb{R}^2 \to \mathbb{R}^2$ satisfies $f(1, 2) = (2, 3)$ and $f(0, 1) = (1, 4)$. Find f, i.e., find $f(x, y)$ for all $x, y \in \mathbb{R}$.

Solution 3.3.7. Let $(x, y) \in \mathbb{R}^2$. Since $\{(1, 2), (0, 1)\}$ forms a basis of $\mathbb{R}^2$, we can express (x, y) as a linear combination:

$$(x, y) = a(1, 2) + b(0, 1) = (a, 2a) + (0, b) = (a, 2a + b).$$

Thus, we have $a = x$ and $b = y - 2x$, giving us $(x, y) = x(1, 2) + (y - 2x)(0, 1)$. Considering the linearity of f, we can write

$$\begin{aligned}
f(x, y) &= f(x(1, 2) + (y - 2x)(0, 1)) \\
&= f(x(1, 2)) + f((y - 2x)(0, 1)) \\
&= xf(1, 2) + (y - 2x)f(0, 1) \\
&= x(2, 3) + (y - 2x)(1, 4) \\
&= (2x, 3x) + (y - 2x, 4y - 8x) \\
&= (2x + y - 2x, 3x + 4y - 8x) \\
&= (y, -5x + 4y).
\end{aligned}$$

Exercise 3.3.8. Let E be a vector space spanned by the vector $u = (1, 1)$. Find a linear mapping f from $\mathbb{R}^2$ to $\mathbb{R}^2$ such that $\ker f = E$.

Solution 3.3.8. The general form of linear maps from $\mathbb{R}^2$ to $\mathbb{R}^2$ is

$$f(x, y) = (ax + by, cx + dy),$$

where $a, b, c, d \in \mathbb{R}$. Since $u \in \ker f$, we get:

$$f(1, 1) = (0, 0) \iff (a + b, c + d) = (0, 0),$$

so $a = -b$ and $c = -d$. Among an infinity of choices, we have $f(x, y) = (x - y, y - x)$, corresponding to $a = 1$, $b = -1$, $c = -1$, and $d = 1$.

Therefore, we can easily see that $\ker f = E$ because

$$\ker f = \{(x, y) \in \mathbb{R}^2 : (x - y, y - x) = (0, 0)\} = \{(x, y) \in \mathbb{R}^2 : x = y\},$$

i.e., $\ker f = \{x(1, 1) : x \in \mathbb{R}\}$.

Exercise 3.3.9. Show that the trace of a matrix is a linear functional.

Solution 3.3.9. We are asked to show that the trace function tr : $M_n(\mathbb{F}) \to \mathbb{F}$ is linear. Let $A = (a_{ij})$ and $B = (b_{ij})$ be both in $M_n(\mathbb{F})$ and let $\alpha \in \mathbb{F}$. Our goal is to verify that $\text{tr}(A + B) = \text{tr}A + \text{tr}B$ and $\text{tr}(\alpha A) = \alpha \text{tr}(A)$.

First, we consider the sum of the matrices $A + B = (a_{ij} + b_{ij})$. We have

$$\text{tr}(A + B) = \sum_{i=1}^{n} (a_{ii} + b_{ii}) = \sum_{i=1}^{n} a_{ii} + \sum_{i=1}^{n} b_{ii} = \text{tr}A + \text{tr}B.$$

Next, we examine the scalar multiplication $\alpha A = (\alpha a_{ij})$. We have

$$\text{tr}(\alpha A) = \sum_{i=1}^{n} (\alpha a_{ii}) = \alpha \sum_{i=1}^{n} a_{ii} = \alpha \text{tr}(A).$$

Thus, we have shown that both properties hold, confirming the linearity of the trace function.

Exercise 3.3.10. Let $f : M_n \to M_n$ be defined by $f(A) = A + A^t$. Show that f is linear.

Solution 3.3.10. The proof is straightforward once we know the basic properties of the transpose of a matrix. So, let $A, B \in M_n$ and α, β be any scalars. We have

$$
\begin{aligned}
f(\alpha A + \beta B) &= \alpha A + \beta B + (\alpha A + \beta B)^t \\
&= \alpha A + \beta B + (\alpha A)^t + (\beta B)^t \\
&= \alpha A + \beta B + \alpha A^t + \beta B^t \\
&= \alpha A + \alpha A^t + \beta B + \beta B^t \text{ (commutativity of + in } M_n) \\
&= \alpha(A + A^t) + \beta(B + B^t) \\
&= \alpha f(A) + \beta f(B),
\end{aligned}
$$

as needed.

Exercise 3.3.11. Let $A, B \in M_n$ be given, and let $T : M_n \to M_n$ be defined by $T(X) = XA - BX$. Show that T is linear.

Solution 3.3.11. Let $X, Y \in M_n$ and α, β be two scalars. By simple properties of multiplication and addition of matrices, we have

$$
\begin{aligned}
T(\alpha X + \beta Y) &= (\alpha X + \beta Y)A - B(\alpha X + \beta Y) \\
&= \alpha XA + \beta YA - \alpha BX - \beta BY \\
&= \alpha XA - \alpha BX + \beta YA - \beta BY \\
&= \alpha(XA - BX) + \beta(YA - BY) \\
&= \alpha T(X) + \beta T(Y),
\end{aligned}
$$

which indicates that T is linear.

Exercise 3.3.12. Let f and g be two linear functionals on E such that $f(x)g(x) = 0$ for all x in E. Show that $f = 0$ or $g = 0$.
Hint: Show that $f(x)g(y) + f(y)g(x) = 0$ for all x, y in E.

Remark. Readers are probably already aware that this result does not apply to a general function. Let us provide a counterexample anyway. Let $f, g : \mathbb{R} \to \mathbb{R}$ be defined by $f(x) = |x| - x$ and $g(x) = |x| + x$ respectively. Despite neither f nor g being the zero function, it is seen that $f(x)g(x) = 0$ for all x in $\mathbb{R}$.

Solution 3.3.12. First, we show the hint. Let x, y be in E. Then $f(x)g(x) = 0$ and $f(y)g(y) = 0$. Since $x + y \in E$, we also have $f(x + y)g(x + y) = 0$. However,

$$
\begin{aligned}
f(x + y)g(x + y) &= [f(x) + f(y)][g(x) + g(y)] \\
&= f(x)g(x) + f(y)g(y) + f(x)g(y) + f(y)g(x) \\
&= f(x)g(y) + f(y)g(x),
\end{aligned}
$$

as wished.

Now, let $(e_k)_k$ be a basis for E. If $f \neq 0$, we know that for at least some k, $f(e_k) \neq 0$. At the same time, $f(e_k)g(e_k) = 0$, which then implies that $g(e_k) = 0$ for all k. Since each x in E may be expressed as $x = \sum_{k=1}^{n} \alpha_k e_k$ (assuming that $\dim E = n$) for some scalars $\alpha_1, \ldots, \alpha_n$, we obtain

$$
g(x) = g\left(\sum_{k=1}^{n} \alpha_k e_k\right) = \sum_{k=1}^{n} \alpha_k g(e_k) = 0,
$$

as wished.

Remark. The solution above is due to Professor Martin Argerami.

Exercise 3.3.13. Let E be a vector space and $f : E \to \mathbb{R}$ be a linear functional that is not identically zero. Show that f is surjective.

Solution 3.3.13. Since f is nonzero, we know that $f(x_0) = y_0 \neq 0$ for some $x_0 \in E$. Now, let $y \in \mathbb{R}$. Then

$$f\left(\frac{yx_0}{y_0}\right) = \frac{y}{y_0}f(x_0) = y,$$

which reveals that f is onto, as needed.

Exercise 3.3.14. Let $A \in M_n$. Show that $\ker(A) = \ker(A^tA)$.

Solution 3.3.14. Let $x \in \ker(A)$, meaning that $Ax = 0$. Hence, $A^t(Ax) = A^t(0) = 0$, which means that $x \in \ker(A^tA)$, thereby $\ker(A) \subset \ker(A^tA)$, and we are halfway through.

Now, suppose $x \in \ker(A^tA)$, that is, $A^tAx = 0$. Thus, $x^tA^tAx = 0$, or equivalently, $(Ax)^tAx = 0$. By Exercise 1.3.46, we can conclude that $Ax = 0$, indicating that $x \in \ker A$. Accordingly, $\ker(A^tA) \subset \ker A$, and so $\ker A = \ker(A^tA)$.

Exercise 3.3.15. Show that the following maps f are linear, find $\ker f$, $\operatorname{ran} f$, their bases and their dimensions, and determine if f is bijective.
  (1) $f : \mathbb{R}^2 \to \mathbb{R}^3$, $f(x, y) = (x - y, x + 3y, y)$.
  (2) $f : \mathbb{R}^3 \to \mathbb{R}^2$, $f(x, y, z) = (2x + z, x - y - z)$.
  (3) $f : \mathbb{R}^3 \to \mathbb{R}^3$, $f(x, y, z) = (x - z, z - y, y - z)$.
  (4) $f : \mathbb{R}^3 \to \mathbb{R}^3$, $f(x, y, z) = (x - y, y, x - y + 2z)$.

Solution 3.3.15. The process for demonstrating the linearity of f is similar each time, so it suffices to show the linearity of the first mapping, for example.

  (1) Let $(x, y), (x', y') \in \mathbb{R}^2$ and let $\alpha, \beta \in \mathbb{R}$. Then,

$$f(\alpha(x, y) + \beta(x', y')) = f(\alpha x + \beta x', \alpha y + \beta y')$$
$$= (\alpha x + \beta x' - (\alpha y + \beta y'), \alpha x + \beta x' + 3(\alpha y + \beta y'), \alpha y + \beta y')$$
$$= (\alpha x - \alpha y, \alpha x + 3\alpha y, \alpha y) + (\beta x' - \beta y', \beta x' + 3\beta y', \beta y')$$
$$= \alpha(x - y, x + 3y, y) + \beta(x' - y', x' + 3y', y')$$
$$= \alpha f(x, y) + \beta f(x', y').$$

Thus, f is linear.
  Next, we have

$$\ker f = \{(x, y) \in \mathbb{R}^2 : f(x, y) = 0_{\mathbb{R}^3} = (0, 0, 0)\}.$$

However, $f(x, y) = (0, 0, 0)$ iff $(x - y, x + 3y, y) = (0, 0, 0)$, which simplifies to

$$\begin{cases} x - y = 0, \\ x + 3y = 0, \\ y = 0, \end{cases}$$

So, $(x, y) = (0, 0)$. Hence, $\ker f = \{(0, 0)\}$, i.e., f is injective, meaning that $\dim \ker f = 0$.

On the other hand, we know that

$$2 = \dim \mathbb{R}^2 = \dim \ker f + \dim \operatorname{ran} f,$$

so $\dim \operatorname{ran} f = 2$. Now we will find a basis of $\operatorname{ran} f$. We can rewrite $\operatorname{ran} f = \{(x - y, x + 3y, y) : x, y \in \mathbb{R}\}$ as $\operatorname{ran} f = \{x \underbrace{(1, 1, 0)}_{u} + y \underbrace{(-1, 3, 1)}_{v} : x, y \in \mathbb{R}\}$. Since $\operatorname{ran} f = \operatorname{span}\{u, v\}$, and $\dim \operatorname{ran} f = 2$, we deduce that $\{u, v\}$ forms a basis of $\operatorname{ran} f$. Thus, f is not surjective because its range is not equal to $\mathbb{R}^3$. In consequence, f is not bijective.

(2) We can start by observing that f is not bijective since

$$\dim \mathbb{R}^3 = 3 \neq \dim \mathbb{R}^2 = 2.$$

Now,

$$\ker f = \{(x, y, z) \in \mathbb{R}^3 : (2x + z, x - y - z) = (0, 0)\}.$$

But,

$$\begin{cases} 2x + z = 0 \\ x - y - z = 0 \end{cases} \iff \begin{cases} z = -2x, \\ y = x - z = 3x. \end{cases}$$

Thus,

$$\ker f = \{(x, 3x, -2x) : x \in \mathbb{R}\} = \{x \underbrace{(1, 3, -2)}_{:=u \neq 0} : x \in \mathbb{R}\},$$

or equivalently, $\ker f = \operatorname{span}\{u\}$, whereby $\dim \ker f = 1$. Hence

$$\dim \operatorname{ran} f = \dim \mathbb{R}^3 - \dim \ker f = 3 - 1 = 2.$$

Also,

$$\begin{aligned} \operatorname{ran} f &= \{(2x + z, x - y - z) : x, y, z \in \mathbb{R}\} \\ &= \{(2x, x) + (0, -y) + (z, -z) : x, y, z \in \mathbb{R}\} \\ &= \{x(2, 1) + y \underbrace{(0, -1)}_{v} + z \underbrace{(1, -1)}_{w} : x, y, z \in \mathbb{R}\}. \end{aligned}$$

So, *three* vectors span ran f, while we already know its dimension is 2! From the calculations for ker f, we easily see that $(2, 1) = -3(0, -1) + 2(1, -1)$. Thus, ran $f = \text{span}\{v, w\}$. Moreover, v and w are linearly independent, and a basis of ran f is then given by $\{v, w\}$.

(3) Since the domain and the codomain have *the same dimension*, we know that f is bijective if and only if ker $f = \{0_{\mathbb{R}^3}\}$. To find ker f, we need to solve the system of equations:

$$\begin{cases} x - z = 0, \\ z - y = 0, \\ y - z = 0, \end{cases}$$

which clearly yields $x = z$ and $y = z$. Hence,

$$\ker f = \{(x, y, z) \in \mathbb{R}^3 : x = z, \ y = z\} = \{(z, z, z) : z \in \mathbb{R}\},$$

which means ker $f = \text{span}\{u\}$, where $u = (1, 1, 1)$, and so dim ker $f = 1$. Thus, f is not injective; therefore, f is not bijective. The dimension of ran f is 2 and readers can easily find that ran f is spanned by $\{(1, 0, 0), (0, -1, 1)\}$, which then forms a basis of ran f.

(4) The transformation f is bijective because the domain and the codomain have the same dimension, namely 3, and ker $f = \{(0, 0, 0)\}$. So, dim ker $f = 0$, and then dim ran $f = 3$. A basis of ran f is given by the set: $\{(1, 0, 1), (-1, 1, -1), (0, 0, 2)\}$.

Exercise 3.3.16. Let $f, g : \mathbb{R}^2 \to \mathbb{R}^2$ be two *linear maps* defined by:
$$f(x, y) = (x, 0) \text{ and } g(x, y) = (x + y, y)$$
Do we have $f \circ g = g \circ f$? What conclusions can we draw from this?

Solution 3.3.16. It is easy to see that $f \circ g, g \circ f : \mathbb{R}^2 \to \mathbb{R}^2$ are given by:

$$(f \circ g)(x) = f(g(x, y)) = f(x + y, y) = (x + y, 0)$$

and

$$(g \circ f)(x, y) = g(f(x, y)) = g(x, 0) = (x, 0)$$

respectively. Thus, we see that $(f \circ g)(x, y) \neq (g \circ f)(x, y)$ for at least some $(x, y) \in \mathbb{R}^2$, indicating that $f \circ g$ is, in general, unequal to $g \circ f$. In another language, (the ring) $(L(\mathbb{R}^2), +, \circ)$ is not commutative.

Exercise 3.3.17. Consider the linear mapping $f : \mathbb{R}^2 \to \mathbb{R}^2$ defined by:
$$f(x, y) = (2y - x, 2x - y).$$
(1) Find the kernel and the range of f.
(2) Determine the invariant points of f and show that they form a vector space in $\mathbb{R}^2$.
(3) Find a complement of the range of f.

Solution 3.3.17.

(1) The kernel of f is trivial, i.e., $\ker f = \{(0, 0)\}$, which implies that $\dim \ker f = 0$. As a result, f is bijective (why?). Thus, $\operatorname{ran} f = \mathbb{R}^2$.

(2) First, recall that (x, y) is an invariant point under f provided $f(x, y) = (x, y)$. In our case, this means that $(2y - x, 2x - y) = (x, y)$, leading to $x = y$. Thus, the set of invariant points is given by
$$E = \{(x, x) : \ x \in \mathbb{R}\} = \{x \underbrace{(1, 1)}_{u} : \ x \in \mathbb{R}\} = \operatorname{span}\{u\},$$
which is a one-dimensional vector space.

(3) We need to find a subspace F such that $E \oplus F = \mathbb{R}^2$. For any (x, y), we can express it as $(x, x) + (0, y - x)$ in a unique way. Therefore, the complement we are looking for is the vector space $F = \{a(0, 1) : \ a \in \mathbb{R}\}$.

Exercise 3.3.18. Let $f : \mathbb{R}^2 \to \mathbb{R}^2$ be a *linear map* defined by:
$$f(x, y) = (x - y, x - 2y).$$
(1) Show that f is injective and explain why it is invertible.
(2) Find f^{-1}.

Solution 3.3.18.

(1) It is clear that $\ker f = \{(0, 0)\}$, so f is injective. Since f is a linear map between vector spaces of equal dimension, f is bijective, so it is invertible as well.

(2) Let (x', y') be such that $f(x, y) = (x - y, x - 2y) = (x', y')$. We want to express (x, y) in terms of (x', y'). We have:
$$\left\{ \begin{array}{l} x - y = x' \\ x - 2y = y' \end{array} \right. \iff \left\{ \begin{array}{l} y = x' - y'. \\ x = 2y + y' = 2x' - y'. \end{array} \right.$$
Hence,
$$f^{-1} : \mathbb{R}^2 \to \mathbb{R}^2, \ f^{-1}(x, y) = (2x - y, x - y).$$

Exercise 3.3.19. Let $f : \mathbb{R}^3 \to \mathbb{R}^3$ be a *linear map* defined by:
$$f(x, y, z) = (3x + y, -2x - 4y + 3z, 5x + 4y - 2z).$$
Prove that f is bijective, the find f^{-1}.

Solution 3.3.19. Readers may quickly check that $\ker f = \{(0, 0, 0)\}$, which is a synonym of f being bijective.

To find the expression of f^{-1}, we need to write (x', y', z') in terms of (x, y, z) given that $f(x, y, z) = (x', y', z')$. Then

$$f(x, y, z) = (x', y', z')$$
$$\Longleftrightarrow (3x + y, -2x - 4y + 3z, 5x + 4y - 2z) = (x', y', z')$$
$$\Longleftrightarrow \begin{cases} 3x + y = x' \\ -2x - 4y + 3z = y' \\ 5x + 4y - 2z = z' \end{cases}$$
$$\Longleftrightarrow \begin{cases} y = x' - 3x \\ -2x - 4(x' - 3x) + 3z = y' \\ 5x + 4(x' - 3x) - 2z = z' \end{cases}$$
$$\Longleftrightarrow \begin{cases} y = x' - 3x \\ 10x + 3z = y' + 4x' \\ -7x - 2z = z' - 4x' \end{cases}$$
$$\Longleftrightarrow \begin{cases} y = x' - 3x \\ 20x + 6z = 2y' + 8x' \\ -21x - 6z = 3z' - 12x' \end{cases}$$
$$\Longleftrightarrow \begin{cases} x = 4x' - 2y' - 3z' \\ y = -11x' + 6y' + 9z' \\ z = -12x' + 7y' + 10z'. \end{cases}$$

Returning to the variables (x, y, z), we can define the inverse mapping $f^{-1} : \mathbb{R}^3 \to \mathbb{R}^3$ as follows:

$$f^{-1}(x, y, z) = (4x - 2y - 3z, -11x + 6y + 9z, -12x + 7y + 10z).$$

Exercise 3.3.20. Let $E = \mathbb{R}_3[X]$ and $f : \mathbb{R}^2 \to E$ be a map defined by
$$f(x, y) = x + y + xX + yX^2 + (x - y)X^3.$$
(1) Prove that f is linear.
(2) Determine the kernel of f and then deduce the rank of f.

Solution 3.3.20.

(1) For any (x, y), $(x', y') \in \mathbb{R}^2$ and any $\alpha, \beta \in \mathbb{R}$,

$$f(\alpha(x, y) + \beta(x', y'))$$
$$= f((\alpha x, \alpha y) + (\beta x', \beta y'))$$
$$= f(\alpha x + \beta x', \alpha y + \beta y')$$
$$= \alpha x + \beta x' + \beta y' + \alpha y + (\alpha x + \beta x')X + (\beta y' + \alpha y)X^2$$
$$+ (\alpha x + \beta x' - \beta y' - \alpha y)X^3$$
$$= \alpha x + \alpha y + \alpha x X + \alpha y X^2 + (\alpha x - \alpha y)X^3$$
$$+ \beta x' + \beta y' + \beta x' X + \beta y' X^2 + (\beta x' - \beta y')X^3$$
$$= \alpha(x + y + xX + yX^2 + (x - y)X^3)$$
$$+ \beta(x' + y' + x'X + y'X^2 + (x' - y')X^3)$$
$$= \alpha f(x, y) + \beta f(x', y').$$

Thus, f is linear.

(2) Recall that $\ker f = \{(x, y) \in \mathbb{R}^2 : f(x, y) = 0_E\}$. Since $f(x, y)$ is a polynomial, it vanishes when all its coefficients vanish, i.e., when

$$\begin{cases} x + y = 0, \\ x = 0, \\ y = 0, \\ x - y = 0. \end{cases}$$

The *unique* solution that satisfies *all* these equations is the null solution, i.e., $(x, y) = (0, 0)$. Whence, $\ker f = \{(0, 0)\}$.

To determine $\dim \operatorname{ran} f$, we can utilize the rank-nullity theorem, which in this context states that

$$\dim \ker f + \dim \operatorname{ran} f = \dim \mathbb{R}^2 = 2.$$

Since $\dim \ker f = 0$, we get $\operatorname{rank} f = \dim \operatorname{ran} f = 2$.

Exercise 3.3.21. Let $E = \mathbb{R}_n[X]$, where $n \in \mathbb{N}$. Define a linear map $f : E \to E$ as follows:

$$f[P(X)] = P(X + 1).$$

Show that f is an isomorphism.

Solution 3.3.21. We will show that f is bijective by showing its invertibility. Let $g : E \to E$ be the map defined by

$$g[P(X)] = P(X - 1).$$

As the linearity of g is trivial, we need only show that $f \circ g = \mathrm{id}_E$. Let $P \in E$. For all $X \in \mathbb{R}$, we can write

$$(f \circ g)[P(X)] = f(g[P(X)]) = f[P(X-1)] = P(X-1+1) = P(X).$$

Consequently, f is invertible, as needed.

Exercise 3.3.22. Let $E = \mathbb{R}_3[X]$, then consider the mapping $f : E \to E$ defined as follows:

$$f(P) = P + (1-X)P'.$$

(1) Check that f is linear.
(2) Find the kernel of f and determine its basis.
(3) Find the range of f and determine its basis.
(4) Show that:

$$\ker f \oplus \operatorname{ran} f = E.$$

Solution 3.3.22.

(1) Once again, the linearity of f is left to the interested readers to explore.

(2) The kernel of f is given by $\ker f = \{P \in E : P + (1-X)P' = 0\}$. Since P has a degree less than or equal to 3, we can express it as $P(X) = aX^3 + bX^2 + cX + d$, where a, b, c, d are real numbers. Hence, $P'(X) = 3aX^2 + 2bX + c$. Thus,

$$P + (1-X)P' = 0$$
$$\Longleftrightarrow aX^3 + bX^2 + cX + d + (1-X)(3aX^2 + 2bX + c) = 0$$
$$\Longleftrightarrow -2aX^3 + (3a - b)X^2 + (2b)X + c + d = 0$$
$$\Longleftrightarrow \begin{cases} -2a = 0 \\ 3a - b = 0 \\ 2b = 0 \\ c + d = 0 \end{cases}$$
$$\Longleftrightarrow \begin{cases} a = 0, \\ b = 0, \\ c = -d. \end{cases}$$

So, $P(X) = c(X-1)$, i.e., $\ker P = \{c(X-1) : c \in \mathbb{R}\}$, which is the vector space spanned by $X - 1$. Thus, its dimension is 1.

(3) The range of f is given by $\operatorname{ran} f = \{P + (1-X)P' : P \in E\}$. Based on the previous calculations, when we replace $P(X)$ with $aX^3 + bX^2 + cX + d$, we get

$$\operatorname{ran} f = \{aX^3 + (3a - b)X^2 + (2b)X + c + d : a, b, c, d \in \mathbb{R}\}.$$

So, $\operatorname{ran} f$ is spanned by *four* vectors, while its dimension is 3 because

$$\dim \operatorname{ran} f = \dim \mathbb{R}_3[X] - \dim \ker f = 4 - 1 = 3.$$

Instead of searching for what is holding us back, notice that since E is spanned by $\{1, X, X^2, X^3\}$, $\operatorname{ran} f$ is spanned by $\{f(1), f(X), f(X^2), f(X^3)\}$. However, $f(1) = 1$, $f(X) = 1$, $f(X^2) = -X^2 + 2X$, and $f(X^3) = -2X^3 + 3X^2$. As $f(1) = f(X)$, we may check whether $\{f(1), f(X^2), f(X^3)\}$ is a free family. Let $a, b, c \in \mathbb{R}$ be such that

$$af(1) + bf(X^2) + cf(X^3) = 0.$$

Then, $a \times 1 + b(-X^2 + 2X) + c(-2X^3 + 3X^2) = 0$, but

$$(-2c)X^3 + (-b + 3c)X^2 + (2b)X + a = 0,$$

which gives $a = b = c = 0$. Since $\dim \operatorname{ran} f = 3$, we deduce that $\{f(1), f(X^2), f(X^3)\}$ is, in effect, a basis of $\operatorname{ran} f$. Therefore,

$$\operatorname{ran} f = \{af(1) + bf(X^2) + cf(X^3) : a, b, c \in \mathbb{R}\}.$$

(4) Among the methods that can be used, we demonstrate that the union of a basis of $\ker f$ and a basis of $\operatorname{ran} f$ creates a basis of $\mathbb{R}_3[X]$. Let us now prove conclusively that $\{X - 1\} \cup \{f(1), f(X^2), f(X^3)\}$ is, in effect, a basis of $\mathbb{R}_3[X]$. We need to demonstrate that its elements are linearly independent. Let $a, b, c, d \in \mathbb{R}$ be such that

$$a(X - 1) + bf(1) + cf(X^2) + df(X^3) = 0,$$

which can be rewritten as

$$a(X - 1) + b + c(-X^2 + 2X) + d(-2X^3 + 3X^2) = 0,$$

and further as

$$(-2d)X^3 + (3d - c)X^2 + (a + 2c)X + b - a = 0.$$

This implies the following system of equations:

$$\begin{cases} -2d = 0, \\ 3d - c = 0, \\ a + 2c = 0, \\ b - a = 0. \end{cases}$$

The unique solution is $(a, b, c, d) = (0, 0, 0, 0)$. Accordingly, $\{X - 1, f(1), f(X^2), f(X^3)\}$ forms a basis of $\mathbb{R}_3[X]$. Consequently,

$$\mathbb{R}_3[X] = \ker f \oplus \operatorname{ran} f.$$

Exercise 3.3.23. Let E and F be two vector spaces and let $f :$ $E \to F$ be a linear transformation. Suppose $\{e_1, \ldots, e_n\}$ is a certain basis of E.

(1) Show that f is injective if and only if the family $\{f(e_1), \ldots, f(e_n)\}$ is free.
(2) Show that f is surjective if and only if the family $\{f(e_1), \ldots, f(e_n)\}$ spans F.
(3) Infer that f is bijective if and only if the family $\{f(e_1), \ldots, f(e_n)\}$ forms a basis of F.

Solution 3.3.23.

(1) If f is injective and $\alpha_1, \ldots, \alpha_n$ are scalars such that $\alpha_1 f(e_1) + \cdots + \alpha_n f(e_n) = 0$, we can rewrite the previous expression as $f(\alpha_1 e_1 + \cdots + \alpha_n e_n) = 0$ due to the linearity of f. Because f is injective, it follows that $\alpha_1 e_1 + \cdots + \alpha_n e_n = 0$. Also, since $f(e_1), \ldots, f(e_n)$ are linearly independent, we conclude that $\alpha_1 = \cdots = \alpha_n = 0$. Therefore, $\{f(e_1), \ldots, f(e_n)\}$ is a linearly independent set.

Conversely, let $\{f(e_1), \ldots, f(e_n)\}$ be free and take any x in $\ker f$. Since x is in E and $\{e_1, \ldots, e_n\}$ is a basis for E, we may write $x = \alpha_1 e_1 + \cdots + \alpha_n e_n$ for some scalars $\alpha_1, \ldots, \alpha_n$. Hence,

$$0 = f(x) = f(\alpha_1 e_1 + \cdots + \alpha_n e_n) = \alpha_1 f(e_1) + \cdots + \alpha_n f(e_n).$$

Therefore, $\alpha_1 = \cdots = \alpha_n = 0$, as the vectors $f(e_1), \ldots, f(e_n)$ are linearly independent. Consequently, $x = 0$, thereby f is injective, as wished.

(2) Suppose f is surjective and let $y \in F$. Then, $y = f(x)$ for some $x \in E$. Also, $x = \alpha_1 e_1 + \cdots + \alpha_n e_n$ for certain scalars $\alpha_1, \ldots, \alpha_n$. Thus,

$$y = f(x) = f(\alpha_1 e_1 + \cdots + \alpha_n e_n) = \alpha_1 f(e_1) + \cdots + \alpha_n f(e_n),$$

which indicates that the family $\{f(e_1), \ldots, f(e_n)\}$ spans F.

Conversely, assume the family $\{f(e_1), \ldots, f(e_n)\}$ spans F. Let $y \in F$. Then, there are scalars $\alpha_1, \ldots, \alpha_n$ such that

$$y = \alpha_1 f(e_1) + \cdots + \alpha_n f(e_n) = f(\alpha_1 e_1 + \cdots + \alpha_n e_n),$$

from which we derive that $y \in \operatorname{ran} f$, or equivalently, f is onto, as needed.

(3) Just apply the previous two properties.

Exercise 3.3.24. Let $f : \mathbb{R}^3 \to \mathbb{R}^3$ be a *linear map* defined by:
$$f(x, y, z) = (0, x, z).$$

(1) Find $\ker f$ and $\operatorname{ran} f$.
(2) Do we have $\mathbb{R}^3 = \ker f \oplus \operatorname{ran} f$?

Solution 3.3.24.

(1) Readers can readily check that
$$\ker f = \{(0, y, 0) : y \in \mathbb{R}\} = \{y(0, 1, 0) : y \in \mathbb{R}\} = \operatorname{span}\{u\},$$

where $u = (0, 1, 0)$. So, $\dim \ker f = 1$, whereby $\dim \operatorname{ran} f = 2$. Besides, $\{(0, 1, 0), (0, 0, 1)\}$ forms a basis of $\operatorname{ran} f$.

(2) The subspaces $\operatorname{ran} f$ and $\ker f$ are not in direct sum because the union of their bases is
$$\{(0, 1, 0), (0, 0, 1), (0, 1, 0)\} = \{(0, 1, 0), (0, 0, 1)\},$$

which cannot be a basis of $\mathbb{R}^3$!

Exercise 3.3.25. (Cf. Exercise 3.4.10) First, recall that a linear mapping $T : F \to F$, where F is a vector space, is called a *projection* provided $T \circ T = T$.

Let $E = \mathbb{R}_2[X]$, and define a map $f : E \to E$ by:
$$f(P) = -\frac{(X+1)^2}{2} P'' + (X+1)P'.$$

(1) Show that f is linear.
(2) Show that f is a projection.
(3) Determine $\ker f$ and $\operatorname{ran} f$, as well as their dimension.
(4) Prove that
$$\mathbb{R}_2[X] = \operatorname{ran} f \oplus \ker f.$$

Solution 3.3.25.

(1) The linearity of f can be easily verified using the basic properties of differentiable functions. We encourage readers to check the details themselves.

(2) Any $P \in E$ may be expressed $P(X) = aX^2 + bX + c$, where $a, b, c \in \mathbb{R}$. Then, $P'(X) = 2aX + b$ and $P''(X) = 2a$. Hence
$$f(P) = f(aX^2 + bX + c) = -\frac{(X+1)^2}{2}(2a) + (X+1)(2aX + b).$$

After simplification, we obtain
$$f(aX^2 + bX + c) = aX^2 + bX - a + b.$$

So,

$$(f \circ f)(P) = f[f(P)] = f(aX^2 + bX - a + b)$$
$$= aX^2 + bX - a + b = f(P),$$

i.e., $f \circ f = f$, meaning that f is a projection.

(3) Writing $P = aX^2 + bX + c$ gives $f(P) = aX^2 + bX - a + b = 0$. Thus, $a = 0$, $b = 0$, and $a = b$, i.e., $a = b = 0$ and $c \in \mathbb{R}$ is arbitrary. Accordingly, $\ker f = \{c\} = \{c \times \mathbf{1}\}$, and so $\dim \ker f = 1$. The latter then yields

$$\mathrm{rank}\, f = \dim \mathrm{ran}\, f = \dim \mathbb{R}_2[X] - \dim \ker f = 3 - 1 = 2.$$

However,

$$\mathrm{ran}\, f = \{aX^2 + bX - a + b : a, b \in \mathbb{R}\}$$
$$= \{a(X^2 - 1) + b(X + 1) : a, b \in \mathbb{R}\},$$

which shows that $\mathrm{ran}\, f$ is spanned by $\{X^2 - 1, X + 1\}$. This is a basis of $\mathrm{ran}\, f$ since its cardinal equals the dimension of $\mathrm{ran}\, f$.

(4) To demonstrate equality, we have various methods available, and from these, we have chosen the following: We have

$$\dim \ker f + \dim \mathrm{ran}\, f = \dim \mathbb{R}_2[X].$$

So, let us show that $\ker f \cap \mathrm{ran}\, f = \{0\}$, or just $\ker f \cap \mathrm{ran}\, f \subset \{0\}$. Assume $P = aX^2 + bX + c \in \ker f \cap \mathrm{ran}\, f$. Then, $P = aX^2 + bX + c \in \ker f$ and $P = aX^2 + bX + c \in \mathrm{ran}\, f$. This leads to $a = b = 0$ and $c = -a + b$, and so $a = b = c = 0$. Thus, $\ker f \cap \mathrm{ran}\, f \subset \{0\}$, thereby $\ker f \cap \mathrm{ran}\, f = \{0\}$, and finally:

$$\mathbb{R}_2[X] = \mathrm{ran}\, f \oplus \ker f.$$

Exercise 3.3.26. Let E be a vector space, let $T : E \to E$ be a projection, and $I : E \to E$ be the identity mapping. Show that $I - T$ remains a projection.

Solution 3.3.26. Let $x \in E$. Then,

$$(I - T)^2 x = (I - T)(I - T)x = (I - T)(x - Tx) = x - Tx - Tx + T^2 x.$$

But, $T^2 = T$, which then means that the previous expression simplifies to $(I - T)x$. Thus, $(I - T)^2 = I - T$, as wished.

Exercise 3.3.27. Let $E = C(\mathbb{R}, \mathbb{R})$, i.e., the vector space of continuous functions from $\mathbb{R}$ to $\mathbb{R}$. We define a map $T : E \to E$ by $Tf(x) = xf(x)$.

    (1) Check that T is linear.

    (2) Find $\ker T$.

    (3) Prove that T is not surjective.

    (4) What does this tell about $\dim E$?

Solution 3.3.27.

(1) Let $f, g \in E$ and let $\alpha, \beta \in \mathbb{R}$. We have for all real x:

$$T(\alpha f + \beta g)(x) = x(\alpha f + \beta g)(x)$$
$$= x\alpha f(x) + x\beta g(x)$$
$$= \alpha x f(x) + \beta x g(x).$$

So, $T(\alpha f + \beta g) = \alpha T(f) + \beta T(g)$, i.e., T is linear.

(2) We claim that T is one-to-one, i.e., $\ker T = \{0_E\}$. It suffices to show that $\ker T \subset \{0_E\}$. Let $f \in \ker T$, which means that $xf(x) = 0$ for all real x, which further implies $f(x) = 0$ for all nonzero real x. However, f is continuous on $\mathbb{R}$, including at zero, and so $f(0) = \lim_{x \to 0} f(x) = 0$. Therefore, $f(x) = 0$ for all $x \in \mathbb{R}$, i.e., $f = 0_E$. As a result, $\ker T = \{0_E\}$, as wished.

(3) The function defined by $g(x) = 1$ for all $x \in \mathbb{R}$, defined on $\mathbb{R}$, is continuous on $\mathbb{R}$. However, if T were onto, we would have for all $x \in \mathbb{R}$: $xf(x) = g(x) = 1$, which is untrue as these equalities are violated if, for instance, $x = 0$. Thus, T is not surjective.

(4) If $\dim E$ were finite, and since f is defined from E to E, we would have an equivalence between the injectivity and the surjectivity of f. Since f is one-to-one without being onto, we must conclude that $\dim E = \infty$.

Exercise 3.3.28. Let $E = \mathbb{R}[X]$ and let $f : E \to E$ be a map defined by $f(P) = P'$, where P' designates the derivative of the polynomial P.

    (1) Prove that f is a surjective linear map.

    (2) Is it injective?

    (3) Infer that the dimension of E is infinite.

Solution 3.3.28.

(1) Let $\alpha, \beta \in \mathbb{R}$ and let $P, Q \in E$. Then

$$f(\alpha P + \beta Q) = (\alpha P + \beta Q)' = \alpha P' + \beta Q' = \alpha f(P) + \beta f(Q),$$

which signifies that f is linear.

To prove that f is surjective, let $Q \in E$. We then have to find at least one P in E such that $f(P) = P' = Q$. If Q is of the form $a_0 + a_1 X + \cdots + a_n X^n$, then we can choose $P = a_0 X + \frac{a_1}{2} X^2 + \cdots + \frac{a_n}{n+1} X^{n+1}$, which is an element of E because it is a polynomial, and its degree is not relevant in this context. It is straightforward to check that $f(P) = Q$, which indicates that f is surjective.

(2) First, recall that a polynomial's derivative is zero if and only if the polynomial is a constant because the derivative of any constant is zero, while the derivative of any non-constant polynomial is nonzero. Due to this, the kernel of f comprises all constant polynomials. In other words, $\ker f = \{c : c \in \mathbb{R}\}$, from which we derive that f is not injective since its kernel is not reduced to $\{0_E\}$.

(3) If we assume that the dimension of E is finite, then according to the rank-nullity theorem, we have:

$$\dim E = \dim \ker f + \dim \operatorname{ran} f.$$

Because $\operatorname{ran} f = E$, as f is surjective, we obtain $\dim \ker f = 0$, which contradicts the earlier finding $\ker f \neq \{0_E\}$. Consequently, the dimension of E is infinite.

Exercise 3.3.29. Let $E = C([0,1], \mathbb{R})$ be the vector space of continuous functions defined from $[0,1]$ to $\mathbb{R}$. Let $\psi : E \to E$ be a map defined by $\psi(f) = T$, where $T(x) = \int_0^x f(t)dt$.
  (1) Show that ψ is an endomorphism of E.
  (2) Show that ψ is injective. Is it surjective?
  (3) Determine $\dim E$.

Solution 3.3.29.

(1) The map ψ is well-defined because T is defined and continuous on $[0,1]$. It is also linear since:

$$\int_0^x (\alpha f(t) + \beta g(t))dt = \alpha \int_0^x f(t)dt + \beta \int_0^x f(t)dt$$

for all f, g in E and all $\alpha, \beta \in \mathbb{R}$. So, ψ is an endomorphism of E

(2) The kernel of ψ is given by:

$$\ker \psi = \left\{ f \in E : T(x) = \int_0^x f(t)dt = 0, \ \forall x \in [0,1] \right\}.$$

If $T(x) = 0$ for all x, then $T'(x)$ (which is $f(x)$) is equal to 0 for all x. This implies that $f = 0$. Therefore, $\ker \psi \subset \{0\}$,

and since the other inclusion is trivial, we have shown that
$\ker \psi = \{0\}$.

Before checking if ψ is surjective, we first note that the
function $x \mapsto T(x) = \int_0^x f(t)dt$ is differentiable on the interval
$[0, 1]$. Now, let $g \in E$. If ψ were surjective, then we would
have:
$$\exists f \in E : \ \psi(f) = g,$$
which means $T(x) = g(x)$ for all x. However, T is differentiable, and g is not necessarily differentiable (for instance, we
can take g to be continuous but non-differentiable). Therefore,
ψ is not surjective.

(3) Suppose that the dimension of E is finite, denoted as $\dim E <
+\infty$. Then,
$$\dim E = \dim \ker \psi + \dim \operatorname{ran} \psi.$$
So,
$$\dim \operatorname{ran} \psi = \dim E - \underbrace{\dim \ker \psi}_{=0} = \dim E.$$

From the above, along with the fact that $\operatorname{ran} \psi \subset E$, we can
see that $\operatorname{ran} \psi = E$, meaning ψ is surjective, which contradicts
Question 2. Thus, $\dim E = +\infty$.

Exercise 3.3.30. Let E be the vector space of functions from $\mathbb{R}$ to
$\mathbb{R}$. We define a map $T : E \to E$ by $Tf(x) = f(x) - f(-x)$.
(1) Check that T is linear.
(2) Find $\ker T$ and $\operatorname{ran} T$.
(3) Is T one-to-one? Onto?

Solution 3.3.30.

(1) Linearity is left to interested readers.
(2) Let $f \in E$ be such that $Tf = 0$, i.e., $f(x) - f(-x) = 0$ for
all $x \in \mathbb{R}$, which means that f is an even function. In other
words, each element in $\ker T$ must be an even function, and
since every even function is clearly in $\ker T$, we conclude by
saying that
$$\ker(T) = \{f : \mathbb{R} \to \mathbb{R} : f \text{ even}\}.$$

We claim that $\operatorname{ran}(T)$ consists of odd functions only, which
is perhaps unexpected. To show this claim, let $g \in \operatorname{ran}(T)$,
which means that $g(x) = f(x) - f(-x)$ for all x in $\mathbb{R}$. Then
$$g(-x) = f(-x) - f(x) = -[f(x) - f(-x)] = -g(x),$$

that is, g is odd. To show the reverse inclusion, let g be an odd function. Writing $g(x) = \frac{g(x)}{2} + \frac{g(x)}{2} = \frac{g(x)}{2} - \frac{g(-x)}{2}$ demonstrates that $g \in \operatorname{ran}(T)$, as needed.

(3) The transformation T is neither injective nor surjective. Indeed, $\ker T \neq \{0\}$ as $f(x) = x^2$, defined from $\mathbb{R}$ into $\mathbb{R}$ is not identically zero, yet it is in $\ker T$. To show that $\operatorname{ran}(T) \neq E$, it is enough to consider any non-odd function, for example, $f(x) = x + 1$ from $\mathbb{R}$ to $\mathbb{R}$. Then $f \in E$ with $f \notin \operatorname{ran}(T)$.

Exercise 3.3.31. Set
$$E = \left\{ A = \begin{pmatrix} a & b \\ b & c \end{pmatrix} : a, b, c \in \mathbb{R} \right\}.$$

(1) Show that E is a subspace of $M_2(\mathbb{R})$ and find a basis of it.
(2) We define the map $f : E \to E$ by
$$f\left[\begin{pmatrix} a & b \\ b & c \end{pmatrix} \right] = \begin{pmatrix} a+c & b \\ b & a+b+c \end{pmatrix}.$$

(a) Prove that f is an endomorphism of E.
(b) Find bases of $\ker f$ and $\operatorname{ran} f$.

Solution 3.3.31.

(1) We leave it to the reader to check that E is a subspace of $M_2(\mathbb{R})$. Its dimension is 3, where a basis of it is $\{M_1, M_2, M_3\}$, where
$$M_1 = \begin{pmatrix} 1 & 0 \\ 0 & 0 \end{pmatrix}, \quad M_2 = \begin{pmatrix} 0 & 1 \\ 1 & 0 \end{pmatrix}, \quad \text{and } M_3 = \begin{pmatrix} 0 & 0 \\ 0 & 1 \end{pmatrix}.$$

(2) (a) Let $\alpha, \beta \in \mathbb{R}$ and $A, B \in E$. Then
$$f(\alpha A + \beta A) = f\left[\alpha \begin{pmatrix} a & b \\ b & c \end{pmatrix} + \beta \begin{pmatrix} a' & b' \\ b' & c' \end{pmatrix} \right]$$
$$= f\left[\begin{pmatrix} \alpha a + \beta a' & \alpha b + \beta b' \\ \alpha b + \beta b' & \alpha c + \beta c' \end{pmatrix} \right]$$
$$= \begin{pmatrix} \alpha a + \beta a' + \alpha c + \beta c' & \alpha b + \beta b' \\ \alpha b + \beta b' & \alpha a + \beta a' + \alpha b + \beta b' + \alpha c + \beta c' \end{pmatrix}$$
$$= \begin{pmatrix} \alpha(a+c) & \alpha b \\ \alpha b & \alpha(a+b+c) \end{pmatrix} + \begin{pmatrix} \beta(a'+c') & \beta b' \\ \beta b' & \beta(a'+b'+c') \end{pmatrix}$$
$$= \alpha \begin{pmatrix} a+c & b \\ b & a+b+c \end{pmatrix} + \beta \begin{pmatrix} a'+c' & b' \\ b' & a'+b'+c' \end{pmatrix}$$
$$= \alpha f(A) + \beta f(B),$$

which means that f is linear.

(b) We have

$$\ker f = \left\{ \begin{pmatrix} a & b \\ b & c \end{pmatrix} : \begin{pmatrix} a+c & b \\ b & a+b+c \end{pmatrix} = \begin{pmatrix} 0 & 0 \\ 0 & 0 \end{pmatrix} \right\}.$$

The solutions of the matrix equation are $b = 0$ and $a = -c$. Therefore,

$$\ker f = \left\{ \begin{pmatrix} a & 0 \\ 0 & -a \end{pmatrix} : a \in \mathbb{R} \right\} = \left\{ a \begin{pmatrix} 1 & 0 \\ 0 & -1 \end{pmatrix} : a \in \mathbb{R} \right\},$$

which is spanned by the nonzero matrix $\begin{pmatrix} 1 & 0 \\ 0 & -1 \end{pmatrix}$. Thus,

$\left\{ \begin{pmatrix} 1 & 0 \\ 0 & -1 \end{pmatrix} \right\}$ is a basis of $\ker f$, thereby $\dim \ker f = 1$.

Before determining the range of f, we know its dimension is 2 because $\dim \operatorname{ran} f = \dim E - \dim \ker f = 3 - 1 = 2$. But,

$$\operatorname{ran} f = \left\{ \begin{pmatrix} a+c & b \\ b & a+b+c \end{pmatrix} : a, b, c \in \mathbb{R} \right\},$$

and so,

$$\operatorname{ran} f = \left\{ (a+c) \begin{pmatrix} 1 & 0 \\ 0 & 1 \end{pmatrix} + b \begin{pmatrix} 0 & 1 \\ 1 & 1 \end{pmatrix} : a, b, c \in \mathbb{R} \right\}.$$

As $\left\{ \begin{pmatrix} 1 & 0 \\ 0 & 1 \end{pmatrix}, \begin{pmatrix} 0 & 1 \\ 1 & 1 \end{pmatrix} \right\}$ spans $\operatorname{ran} f$ and $\dim \operatorname{ran} f = 2$, we conclude that it is a basis of $\operatorname{ran} f$.

Exercise 3.3.32. Let $A, B \in M_n$ be given and define $T : M_n \to M_n$ by $T(X) = AXB$.
  (1) Show that T is linear.
  (2) Show that T is bijective if A and B are both invertible.

Solution 3.3.32.
  (1) Let $X, Y \in M_n$ and let α, β be two scalars. We have

$$T(\alpha X + \beta Y) = A(\alpha X + \beta Y)B = (\alpha AX + \beta AY)B$$
$$= \alpha AXB + \beta AYB,$$

meaning that $T(\alpha X + \beta Y) = \alpha T(X) + \beta T(Y)$, indicating that T is linear.
  (2) Since M_n is a finite-dimensional vector space, to show that T is bijective; it suffices to show that T is injective (see Theorem 3.1.3, and a proof of this result appears in Exercise 3.3.34). So, let X be such that $T(X) = 0_{M_n}$, i.e., $AXB = 0_{M_n}$. Since

A and B are invertible, we have $X = A^{-1}AXBB^{-1} = 0_{M_n}$, showing that $\ker T \subseteq \{0_{M_n}\}$, and since the other inclusion is always true (why?), we conclude that $\ker T = \{0_{M_n}\}$, which is a synonym of the injectivity of T.

Exercise 3.3.33. Find a linear transformation that is right invertible without being left invertible.

Recall that a linear map $f : E \to E$ is right (resp. left) invertible if there exists a linear $g \in L(E)$ such that $g \circ f = I$ (resp. $f \circ g = I$), where $I : E \to E$ is the identity map on E.

Solution 3.3.33. Let $E = \mathbb{R}[X]$ and $f : E \to E$ be a map defined by $f(P) = P'$, where P' designates the derivative of the polynomial P. Consider $g : E \to E$ defined by $g(P)(x) = \int_0^x P(t)dt$. Then, for all $P \in E$

$$f[g(P)(x)] = \left(\int_0^x P(t)dt \right)' = P(X),$$

that is, $f \circ g = I$. In other words, f is right invertible.

Next, we want to show that $g \circ f \neq I$, meaning that f is not left invertible. By the Fundamental Theorem of Calculus,

$$(g \circ f)(P)(x) = \int_0^x P'(t)dt = P(X) - P(0),$$

which is not equal to $P(X)$ unless $P(0) = 0$. Therefore, $g \circ f \neq I$ in general.

Exercise 3.3.34. Let $f : E \to E$ be a linear map and $\dim E < +\infty$. Show that

$$f \text{ injective} \iff f \text{ surjective} \iff f \text{ bijective}.$$

Deduce that if $f : E \to E$ is left invertible (resp. right invertible), then f is invertible. Finally, interpret this result in the context of matrices.

Solution 3.3.34. It is clear that we need only establish the equivalence between injectivity and surjectivity. If f is injective, this means that $\ker f = \{0\}$, and so $\dim \ker f = 0$. Thus, by Theorem 3.1.4 we have $\dim \operatorname{ran} f = \dim E$, which, and since $\operatorname{ran} f$ is a subspace of E, leads to $\operatorname{ran} f = E$, i.e., f is onto.

Conversely, suppose f is surjective, that is, $\operatorname{ran} f = E$. So, Theorem 3.1.4 yields $\dim \ker f = 0$, which forces $\ker f = \{0\}$, which is a synonym of the injectivity of f, as required.

Suppose $f : E \to E$ is left invertible, meaning there exists a linear map $g : E \to E$ such that $g \circ f = I$, where I is the identity map on E.

This implies that f is injective. To see this, assume $f(x) = 0$ for some $x \in E$. Applying g to both sides gives:

$$g(f(x)) = g(0) = 0.$$

Since $g \circ f = I$, we have $g(f(x)) = x$, so $x = 0$. Thus, f is injective.

Next, by the previous question, we know that f must be surjective. Therefore, f is bijective, and hence invertible.

A similar argument holds if f is right invertible. The proof in that case follows the same steps, and is left to the interested reader.

Finally, let A and B be two square matrices associated with f and g respectively, e.g., in the case $f \circ g = I$. This means that $AB = I$, i.e., A is right invertible. By the previous results, A is invertible.

Exercise 3.3.35. Write the matrix associated with the linear map $f : M_2(\mathbb{R}) \to \mathbb{R}$ defined by $f(A) = \mathrm{tr}A$, with respect to the canonical bases.

Solution 3.3.35. Denote the matrix associated with f, which is of size 1×2^2, by T. Recall that the standard basis of $M_2(\mathbb{R})$ is given by

$$\left\{ E_{11} = \begin{pmatrix} 1 & 0 \\ 0 & 0 \end{pmatrix}, \ E_{12} = \begin{pmatrix} 0 & 1 \\ 0 & 0 \end{pmatrix}, \ E_{21} = \begin{pmatrix} 0 & 0 \\ 1 & 0 \end{pmatrix}, \ E_{22} = \begin{pmatrix} 0 & 0 \\ 0 & 1 \end{pmatrix} \right\}.$$

Since $\mathrm{tr}E_{11} = \mathrm{tr}E_{22} = 1$ and $\mathrm{tr}E_{12} = \mathrm{tr}E_{21} = 0$, the matrix T is given by:

$$T = \begin{pmatrix} 1 & 0 & 0 & 1 \end{pmatrix}.$$

Exercise 3.3.36. Let $f : \mathbb{R} \to \mathbb{R}^2$, $g : \mathbb{R}^2 \to \mathbb{R}^2$ be two linear mappings defined by:

$$f(x) = (-x, x) \text{ and } g(x, y) = (x + y, x - 3y)$$

respectively.

(1) Find $g \circ f$.
(2) Find the matrices associated with f, g and $g \circ f$ with respect to their canonical bases.

Solution 3.3.36.

(1) The composition $g \circ f$ is defined from $\mathbb{R}$ to $\mathbb{R}^2$ by:

$$(g \circ f)(x) = g(f(x)) = g(-x, x) = (0, -4x).$$

(2) The canonical basis of $\mathbb{R}$ is $\{1\}$. Then, we have $f(1) = (-1, 1)$, which signifies that the associated matrix is given by:

$$M_f = \begin{pmatrix} -1 \\ 1 \end{pmatrix}.$$

As for g, we have $g(1,0) = (1,1)$ and $g(0,1) = (1,-3)$, and so

$$M_g = \begin{pmatrix} 1 & 1 \\ 1 & -3 \end{pmatrix}.$$

To find $M_{g \circ f}$, we will present two methods to achieve this.

(a) **First method:** Since $(g \circ f)(1) = (0,-4)$, we see that

$$M_{g \circ f} = \begin{pmatrix} 0 \\ -4 \end{pmatrix}.$$

(b) **Second method:** We know that $M_{g \circ f} = M_g M_f$, and so

$$M_{g \circ f} = \begin{pmatrix} 1 & 1 \\ 1 & -3 \end{pmatrix} \begin{pmatrix} -1 \\ 1 \end{pmatrix} = \begin{pmatrix} 0 \\ -4 \end{pmatrix}.$$

Exercise 3.3.37. Let $E = \mathbb{R}_3[X]$ and let $f : \mathbb{R}^2 \to E$ be a *linear* map defined by:

$$f(x,y) = x + y + xX + yX^2 + (x-y)X^3.$$

(1) Determine the matrix associated with f, noted A, with respect to the canonical bases of $\mathbb{R}^2$ and E.

(2) Let $g : E \to \mathbb{R}^2$ be defined by:

$$g[P(X)] = (P(0), P(1)).$$

Say why $g \circ f$ is linear and give its matrix in the canonical basis of $\mathbb{R}^2$.

Solution 3.3.37.

(1) Since $\dim E = 4$ and $\dim \mathbb{R}^2 = 2$, matrix A is of size 4×2. We have

$$f(1,0) = 1 + X + X^3 = \mathbf{1} \times 1 + \mathbf{1} \times X + \mathbf{0} \times X^2 + \mathbf{1} \times X^3$$

and

$$f(0,1) = 1 + X^2 - X^3 = \mathbf{1} \times 1 + \mathbf{0} \times X + \mathbf{1} \times X^2 + (\mathbf{-1}) \times X^3.$$

Therefore,

$$A = \begin{pmatrix} 1 & 1 \\ 1 & 0 \\ 0 & 1 \\ 1 & -1 \end{pmatrix}.$$

(2) We can easily show that g is linear, then $g \circ f$ be linear (since f is already linear). To find the matrix C associated with $g \circ f$, we can either multiply the matrix of g by the matrix of f or find it directly. Let us work through both methods

(a) **First method:** The matrix associated with g is of size 2×4, denoted by B. Since

$$g(1) = (1,1), \ g(X) = (0,1), \ g(X^2) = (0,1)$$

and

$$g(X^3) = (0,1),$$

it follows that

$$B = \begin{pmatrix} 1 & 0 & 0 & 0 \\ 1 & 1 & 1 & 1 \end{pmatrix}.$$

So,

$$C = BA = \begin{pmatrix} 1 & 0 & 0 & 0 \\ 1 & 1 & 1 & 1 \end{pmatrix} \begin{pmatrix} 1 & 1 \\ 1 & 0 \\ 0 & 1 \\ 1 & -1 \end{pmatrix} = \begin{pmatrix} 1 & 1 \\ 3 & 1 \end{pmatrix}.$$

(b) **Second method:** To begin, we explicitly find $g \circ f$. Let $(x, y) \in \mathbb{R}^2$. Then

$$(g \circ f)(x, y) = g[f(x, y)]$$
$$= g(x + y + xX + yX^2 + (x - y)X^3),$$

which simplifies to

$$(g \circ f)(x, y) = (x + y, x + y + x + y + x - y)$$
$$= (x + y, 3x + y).$$

Since

$$(g \circ f)(1, 0) = (1, 3) \text{ and } (g \circ f)(0, 1) = (1, 1),$$

it ensues that

$$C = \begin{pmatrix} 1 & 1 \\ 3 & 1 \end{pmatrix}.$$

Exercise 3.3.38. Let $f : \mathbb{R}^2 \to \mathbb{R}^2$ be a *linear* map defined by:

$$f(x, y) = (x - 2y, 3y).$$

(1) Find the matrix A associated with f with respect to the canonical basis of $\mathbb{R}^2$, denoted by $\{e_1, e_2\}$.
(2) Let $e_1' = e_1$ and $e_2' = e_1 + e_2$. Show that $\{e_1', e_2'\}$ is a new basis of $\mathbb{R}^2$.
(3) Using two methods, find the matrix of f in this new basis.

Solution 3.3.38.

(1) Since

$$f(e_1) = f(1, 0) = (1 - 2 \times 0, 3 \times 0) = (1, 0)$$

and

$$f(e_2) = f(0, 1) = (0 - 2, 3) = (-2, 3),$$

it follows that

$$A = \begin{pmatrix} 1 & -2 \\ 0 & 3 \end{pmatrix}.$$

(2) Given that card$\{e'_1, e'_2\} = 2 = \dim \mathbb{R}^2$, for $\{e'_1, e'_2\}$ to be a basis, we need only show it is free. However, the vectors $e'_1 = e_1 = (1,0)$ and $e'_2 = e_1 + e_2 = (1,1)$ are not proportional, i.e., they are linearly independent. Thus, $\{e'_1, e'_2\}$ is a basis of $\mathbb{R}^2$.

(3) (a) **First method:** The transition matrix from $\{e_1, e_2\}$ to $\{e'_1, e'_2\}$ is given by:

$$P = \begin{pmatrix} 1 & 1 \\ 0 & 1 \end{pmatrix}.$$

The matrix of f in this new basis, denoted by B, is given by the formula $B = P^{-1}AP$. So, we need to find P^{-1}. We can use one of the methods from previous exercises to find that $P^{-1} = \begin{pmatrix} 1 & -1 \\ 0 & 1 \end{pmatrix}$. Thus,

$$B = \begin{pmatrix} 1 & -1 \\ 0 & 1 \end{pmatrix} \begin{pmatrix} 1 & -2 \\ 0 & 3 \end{pmatrix} \begin{pmatrix} 1 & 1 \\ 0 & 1 \end{pmatrix},$$

so that

$$B = \begin{pmatrix} 1 & -4 \\ 0 & 3 \end{pmatrix}.$$

(b) **Second method:** To find the matrix of f with respect to the new basis $\{e'_1, e'_2\}$, which we denote by B, we compute $f(e'_1)$ and $f(e'_2)$ in terms of e'_1 and e'_2. We have

$$f(e'_1) = f(e_1) = (1,0) = e'_1$$

and

$$f(e'_2) = f(e_1 + e_2) = f(e_1) + f(e_2)$$
$$= (1,0) + (-2,3) = (-1,3).$$

But,

$$(-1,3) = -e_1 + 3e_2 = -e'_1 + 3(e'_2 - e'_1) = -4e'_1 + 3e'_2.$$

Thus,

$$B = \begin{pmatrix} 1 & -4 \\ 0 & 3 \end{pmatrix}.$$

Exercise 3.3.39. Let $f : \mathbb{R}^3 \to \mathbb{R}^3$ be a linear map defined by:
$$f(x, y, z) = (-2z, x + 2y + z, x + 3z).$$
(1) Find the matrix associated with f in the canonical basis of $\mathbb{R}^3$, noted $\{e_1, e_2, e_3\}$.
(2) Let:
$$e_1' = -e_1 + e_3, \quad e_2' = e_2, \quad e_3' = -2e_1 + e_2 + e_3.$$
Prove that $\{e_1', e_2', e_3'\}$ is another basis of $\mathbb{R}^3$.
(3) Check that the matrix associated with f with respect to the basis $\{e_1', e_2', e_3'\}$ is diagonal.

Solution 3.3.39. We will only include a few details.

(1) Let A be such a matrix. Then
$$A = \begin{pmatrix} 0 & 0 & -2 \\ 1 & 2 & 1 \\ 1 & 0 & 3 \end{pmatrix}.$$

(2) Showing that $\{e_1', e_2', e_3'\}$ forms a linearly independent set proves that it is a basis, which readers can verify quickly.
(3) The transition matrix from $\{e_1, e_2, e_3\}$ to $\{e_1', e_2', e_3'\}$ is given by:
$$P = \begin{pmatrix} -1 & 0 & -2 \\ 0 & 1 & 1 \\ 1 & 0 & 1 \end{pmatrix}$$

and its inverse is give by:
$$P^{-1} = \begin{pmatrix} 1 & 0 & 2 \\ 1 & 1 & 1 \\ -1 & 0 & -1 \end{pmatrix}.$$

Thus, the matrix associated with f in the new basis $\{e_1', e_2', e_3'\}$, denoted by B, is given by:
$$B = P^{-1}AP = \begin{pmatrix} 1 & 0 & 2 \\ 1 & 1 & 1 \\ -1 & 0 & -1 \end{pmatrix} \begin{pmatrix} 0 & 0 & -2 \\ 1 & 2 & 1 \\ 1 & 0 & 3 \end{pmatrix} \begin{pmatrix} -1 & 0 & -2 \\ 0 & 1 & 1 \\ 1 & 0 & 1 \end{pmatrix}$$
$$= \begin{pmatrix} 2 & 0 & 0 \\ 0 & 2 & 0 \\ 0 & 0 & 1 \end{pmatrix},$$

which is obviously diagonal.

Exercise 3.3.40. Consider the matrix:

$$A = \begin{pmatrix} 2 & -1 & 1 \\ 3 & 2 & -3 \end{pmatrix}.$$

Let $f : \mathbb{R}^3 \to \mathbb{R}^2$ be the linear map associated with matrix A with respect to the standard bases of $\mathbb{R}^3$ and $\mathbb{R}^2$, denoted by $\{e_1, e_2, e_3\}$ and $\{f_1, f_2\}$ respectively. Let:

$$\begin{cases} e'_1 = e_2 + e_3, \\ e'_2 = e_1 + e_3, \\ e'_3 = e_1 + e_2, \end{cases} \quad \text{and} \quad \begin{cases} f'_1 = \frac{1}{2}(f_1 + f_2), \\ f'_2 = \frac{1}{2}(f_1 - f_2). \end{cases}$$

(1) Show that $\{e'_1, e'_2, e'_3\}$ is a basis of $\mathbb{R}^3$ and that $\{f'_1, f'_2\}$ is a basis of $\mathbb{R}^2$.

(2) Give the matrix associated with f in this new basis.

Solution 3.3.40.

(1) Readers may verify for themselves that $\{e'_1, e'_2, e'_3\}$ and $\{f'_1, f'_2\}$ are bases of $\mathbb{R}^3$ and $\mathbb{R}^2$, respectively.

(2) The transition matrix of $\{e_1, e_2, e_3\}$ to $\{e'_1, e'_2, e'_3\}$ is given by:

$$P = \begin{pmatrix} 0 & 1 & 1 \\ 1 & 0 & 1 \\ 1 & 1 & 0 \end{pmatrix}.$$

Also, the transition matrix from $\{f_1, f_2\}$ to $\{f'_1, f'_2\}$ is given by:

$$Q = \frac{1}{2} \begin{pmatrix} 1 & 1 \\ 1 & -1 \end{pmatrix}.$$

We also know that matrix B in the new basis is given by the formula $B = Q^{-1}AP$. From different ways to find Q^{-1}, here we choose to express the elements of the basis $\{f_1, f_2\}$ in terms of the basis $\{f'_1, f'_2\}$, which allows us to determine Q^{-1}. We have

$$\begin{cases} f'_1 = \frac{1}{2}(f_1 + f_2), \\ f'_2 = \frac{1}{2}(f_1 - f_2), \end{cases} \Longleftrightarrow \begin{cases} 2f'_1 = f_1 + f_2, \\ 2f'_2 = f_1 - f_2, \end{cases}$$

$$\Longleftrightarrow \begin{cases} f_1 = f'_1 + f'_2, \\ f_2 = f'_1 - f'_2. \end{cases}$$

So,

$$Q^{-1} = \begin{pmatrix} 1 & 1 \\ 1 & -1 \end{pmatrix}.$$

Thus,

$$B = Q^{-1}AP = \begin{pmatrix} 1 & 1 \\ 1 & -1 \end{pmatrix} \begin{pmatrix} 2 & -1 & 1 \\ 3 & 2 & -3 \end{pmatrix} \begin{pmatrix} 0 & 1 & 1 \\ 1 & 0 & 1 \\ 1 & 1 & 0 \end{pmatrix}.$$

Therefore,

$$B = \begin{pmatrix} -1 & 3 & 6 \\ 1 & 3 & -4 \end{pmatrix}.$$

Exercise 3.3.41. Let $E = \mathbb{R}_2[X]$ and $f : E \to E$ be the linear map defined by $f[P(X)] = P(X+1)$.
   (1) Determine the matrix A associated with f with respect to the standard basis of $\mathbb{R}_2[X]$.
   (2) Prove that $\{1, X, \frac{1}{2}(X^2 - X)\}$ is a new basis of $\mathbb{R}_2[X]$.
   (3) Determine the matrix B associated with f in the new basis.

Solution 3.3.41.

   (1) Since

$$f(1) = 1 = \mathbf{1} \times 1 + \mathbf{0} \times X + \mathbf{0} \times X^2,$$

$$f(X) = X + 1 = \mathbf{1} \times 1 + \mathbf{1} \times X + \mathbf{0} \times X^2,$$

and

$$f(X^2) = (X+1)^2 = X^2 + 2X + 1 = \mathbf{1} \times 1 + \mathbf{2} \times X + \mathbf{1} \times X^2,$$

it ensues that

$$A = \begin{pmatrix} 1 & 1 & 1 \\ 0 & 1 & 2 \\ 0 & 0 & 1 \end{pmatrix}.$$

   (2) Since $\text{card}\{1, X, \frac{1}{2}(X^2 - X)\} = 3 = \dim \mathbb{R}_2[X]$, we know that $\{1, X, \frac{1}{2}(X^2 - X)\}$ is a basis of $\mathbb{R}_2[X]$ once its elements are linearly independent. Let $\alpha, \beta, \gamma \in \mathbb{R}$ be such that $\alpha + \beta X + \gamma\frac{1}{2}(X^2 - X) = 0$. Hence,

$$\alpha + (\beta - \frac{\gamma}{2})X + \frac{\gamma}{2}X^2 = 0,$$

but a polynomial vanishes if all its coefficients vanish, i.e., when $\alpha = \beta = \gamma = 0$. Therefore, $\{1, X, \frac{1}{2}(X^2 - X)\}$ is a basis of $\mathbb{R}_2[X]$.

(3) Since $1 = \mathbf{1} \times 1 + \mathbf{0} \times X + \mathbf{0} \times X^2$, $X = \mathbf{0} \times 1 + \mathbf{1} \times X + \mathbf{0} \times X^2$ and $\frac{1}{2}(X^2 - X) = \mathbf{0} \times 1 - \frac{1}{2} \times X + \frac{1}{2} \times X^2$, the transition matrix from $\{1, X, X^2\}$ to $\{1, X, \frac{1}{2}(X^2 - X)\}$ is given by:

$$P = \begin{pmatrix} 1 & 0 & 0 \\ 0 & 1 & -\frac{1}{2} \\ 0 & 0 & \frac{1}{2} \end{pmatrix},$$

while its inverse is given by:

$$P^{-1} = \begin{pmatrix} 1 & 0 & 0 \\ 0 & 1 & 1 \\ 0 & 0 & 2 \end{pmatrix}.$$

Consequently,

$$B = P^{-1}AP = \begin{pmatrix} 1 & 1 & 0 \\ 0 & 1 & 1 \\ 0 & 0 & 1 \end{pmatrix}.$$

Exercise 3.3.42.

(1) Let f be a linear transformation from $\mathbb{R}^3$ to $\mathbb{R}^3$ with a matrix representation in the canonical basis given by:

$$A = \begin{pmatrix} 3 & -1 & 1 \\ 0 & 2 & 0 \\ 1 & -1 & 3 \end{pmatrix}.$$

Give the expression of the matrix B, which represents the matrix A in the new basis of $\mathbb{R}^3$, formed by the vectors:

$$e'_1 = e_1 - e_3, \ e'_2 = e_2 + e_3, \ e'_3 = e_1 + e_3.$$

(2) Let $n \in \mathbb{N}$. Find B^n and then deduce A^n.

(3) Now, consider three sequences (x_n), (y_n), and (z_n) defined by:

$$\begin{cases} x_0 = y_0 = 1, \ z_0 = 2, \\ x_{n+1} = 3x_n - y_n + z_n, \\ y_{n+1} = 2y_n, \\ z_{n+1} = x_n - y_n + 3z_n. \end{cases}$$

Find the general terms of x_n, y_n, and z_n.

Solution 3.3.42.

(1) It is left to the readers to confirm that:

$$B = P^{-1}AP = \begin{pmatrix} 2 & 0 & 0 \\ 0 & 2 & 0 \\ 0 & 0 & 4 \end{pmatrix},$$

where the transition matrix P and its inverse are given by:

$$P = \begin{pmatrix} 1 & 0 & 1 \\ 0 & 1 & 0 \\ -1 & 1 & 1 \end{pmatrix} \text{ and } P^{-1} = \frac{1}{2} \begin{pmatrix} 1 & 1 & -1 \\ 0 & 2 & 0 \\ 1 & -1 & 1 \end{pmatrix},$$

respectively.

(2) Since B is diagonal, we have for all $n \in \mathbb{N}$

$$B^n = \begin{pmatrix} 2^n & 0 & 0 \\ 0 & 2^n & 0 \\ 0 & 0 & 4^n \end{pmatrix}.$$

Since $A = PBP^{-1}$, we can write

$$\begin{aligned} A^n &= (PBP^{-1})^n \\ &= (PBP^{-1})(PBP^{-1})\cdots(PBP^{-1}) \\ &= PB(P^{-1}P)B(P^{-1}P)\cdots(P^{-1}P)BP^{-1} \\ &= PB^nP^{-1}. \end{aligned}$$

After performing calculations, we find that:

$$A^n = 2^{n-1} \begin{pmatrix} 1+2^n & 1-2^n & 2^n-1 \\ 0 & 2 & 0 \\ 2^n-1 & 1-2^n & 1+2^n \end{pmatrix}.$$

(3) The given system is equivalent to $X_{n+1} = AX_n$, for all $n \in \mathbb{N}$, where $X_n = \begin{pmatrix} x_n \\ y_n \\ z_n \end{pmatrix}$. By induction, we have

$$X_n = AX_{n-1} = A(AX_{n-2}) = A^2 X_{n-2} = \cdots = A^n X_0$$

for all n. Thus,

$$X_n = \begin{pmatrix} x_n \\ y_n \\ z_n \end{pmatrix} = 2^{n-1} \begin{pmatrix} 1+2^n & 1-2^n & 2^n-1 \\ 0 & 2 & 0 \\ 2^n-1 & 1-2^n & 1+2^n \end{pmatrix} \begin{pmatrix} x_0 \\ y_0 \\ z_0 \end{pmatrix},$$

which then yields

$$\begin{cases} x_n = 2^{2n}, \\ y_n = 2^n, \\ z_n = 2^n(1+2^n), \end{cases}$$

still for all n.

Exercise 3.3.43. Let $\{e^{-x}, xe^{-x}, x^2e^{-x}\}$ be a set of functions from $\mathbb{R}$ to $\mathbb{R}$. We set $E = \text{span}\{e^{-x}, xe^{-x}, x^2e^{-x}\}$.

(1) Find $\dim E$.

(2) Define $T : E \to E$ by $T(f) = f'$. Show that T is an endomorphism of E. Determine its matrix A in the basis: $\{e^{-x}, xe^{-x}, x^2e^{-x}\}$.

(3) Let:

$$B = \begin{pmatrix} 0 & 1 & 0 \\ 0 & 0 & 2 \\ 0 & 0 & 0 \end{pmatrix} \text{ and } I = \begin{pmatrix} 1 & 0 & 0 \\ 0 & 1 & 0 \\ 0 & 0 & 1 \end{pmatrix}.$$

(a) Compute B^n for each n in $\mathbb{N}$.

(b) Deduce that:

$$A^n = (-1)^n \left(I - nB + \frac{n(n-1)}{2}B^2 \right), \quad \forall n \in \mathbb{N}.$$

(4) Let $n \in \mathbb{N}$. Using the matrix A^n, find the nth derivative of the function $g : \mathbb{R} \to \mathbb{R}$ defined by:

$$g(x) = (3 - 2x + 8x^2)e^{-x}.$$

Solution 3.3.43.

(1) As E is spanned by the functions e^{-x}, xe^{-x}, and x^2e^{-x}, its dimension will be 3 once we establish the linear independence of its elements. Let $a, b, c \in \mathbb{R}$ be such that

$$ae^{-x} + bxe^{-x} + cx^2e^{-x} = 0, \ \forall x \in \mathbb{R},$$

which simplifies to

$$a + bx + cx^2 = 0, \ \forall x \in \mathbb{R}.$$

Thus, $a = b = c = 0$, and the set $\{e^{-x}, xe^{-x}, x^2e^{-x}\}$ is a basis of E, thereby $\dim E = 3$.

(2) We can easily check that T is linear from E to E, which is then an endomorphism of E. To find the matrix associated with T, we evaluate (for all x):

$$T(e^{-x}) = -e^{-x} = \mathbf{-1} \times e^{-x} + \mathbf{0}xe^{-x} + \mathbf{0}x^2e^{-x},$$

$$T(xe^{-x}) = e^{-x} - xe^{-x} = \mathbf{1} \times e^{-x} - \mathbf{1} \times xe^{-x} + \mathbf{0}x^2e^{-x},$$

and

$$T(x^2e^{-x}) = 2xe^{-x} - x^2e^{-x} = \mathbf{0} \times e^{-x} + \mathbf{2} \times xe^{-x} - \mathbf{1} \times x^2e^{-x}.$$

Hence,

$$A = \begin{pmatrix} -1 & 1 & 0 \\ 0 & -1 & 2 \\ 0 & 0 & -1 \end{pmatrix}.$$

(3) (a) It is easy to see that

$$B^2 = \begin{pmatrix} 0 & 0 & 2 \\ 0 & 0 & 0 \\ 0 & 0 & 0 \end{pmatrix} \text{ and } B^3 = \begin{pmatrix} 0 & 0 & 0 \\ 0 & 0 & 0 \\ 0 & 0 & 0 \end{pmatrix}.$$

So, $B^n = 0_{M_3(\mathbb{R})}$ for all $n \geq 3$.

(b) First, observe that $A = -I + B$. We have two possible approaches: Applying the binomial formula for I and B or proving the result by induction, given that we already know the expected form of A^n. However, we opt for an inductive proof. The base case follows from the identity

$$A^2 = I - 2IB + B^2,$$

which holds due to the commutativity of I and B. Now, assume the formula for A^n holds. Then, and as $B^3 = 0_{M_3}(\mathbb{R})$,

$$A^{n+1} = A^n A$$

$$= (-1)^n \left(I - nB + \frac{n(n-1)}{2}B^2 \right)(-I + B)$$

$$= (-1)^n \left(-I + nB - \frac{n(n-1)}{2}B^2 + B - nB^2 \right)$$

$$= (-1)^{n+1} \left(I - (n+1)B + \frac{n(n+1)}{2}B^2 \right),$$

Thus, we have shown the formula for A^n. The matrix A^n is then written for all n as:

$$A^n = (-1)^n \begin{pmatrix} 1 & -n & n(n-1) \\ 0 & 1 & -2n \\ 0 & 0 & 1 \end{pmatrix}.$$

(4) First, we should refrain from using the Leibniz formula, even though it is applicable to the functions $x \mapsto e^{-x}$ and $x \mapsto 3 - 2x + 8x^2$, since another method has been specified. Now, we can write:

$$g' = T(g), \ g'' = T(g') = (T \circ T)(g), \dots,$$

and

$$g^{(n)} = (\underbrace{T \circ T \circ \cdots \circ T}_{n \text{ times}})(g).$$

So, the matrix of T^n corresponds to A^n. Thus, the matrix of $g^{(n)}$ is given by:

$$(-1)^n \begin{pmatrix} 1 & -n & n(n-1) \\ 0 & 1 & -2n \\ 0 & 0 & 1 \end{pmatrix} \begin{pmatrix} 3 \\ -2 \\ 8 \end{pmatrix} = (-1)^n \begin{pmatrix} 8n^2 - 6n + 3 \\ -16n - 2 \\ 8 \end{pmatrix}.$$

Accordingly,

$$g^{(n)}(x) = (-1)^n e^{-x} (8n^2 - 6n + 3 - (16n + 2)x + 8x^2)$$

for $n = 1, 2, \ldots$

3.4. Exercises without Solutions

Exercise 3.4.1. Give the general form of linear maps from $\mathbb{R}^2$ to $\mathbb{R}^2$. What is the general form of linear maps from $\mathbb{R}^3$ to $\mathbb{R}^3$ (cf. Exercise 3.3.4)?

Exercise 3.4.2. Identify which of the following maps are linear and provide justification for your answer.

(1) $f : \mathbb{Z} \times \mathbb{Z} \to \mathbb{R}^2$, $f(x, y) = (2x - y, -3x + 2y)$.
(2) $f : \mathbb{R}^2 \to \mathbb{R}^2$, $f(x, y) = (|x|, x + 2y)$.
(3) $f : \mathbb{R}^3 \to \mathbb{R}^3$, $f(x, y, z) = (0, 0, 0)$.
(4) $f : \mathbb{R}^3 \to \mathbb{R}^3$, $f(x, y, z) = (z, y, x)$.
(5) $f : \mathbb{R}^3 \to \mathbb{R}^4$, $f(x, y, z) = (x + y, x - z, z + 1, 0)$.

Exercise 3.4.3. Find the linear map $f : \mathbb{R}^2 \to \mathbb{R}^2$ that satisfies $f(1, 2) = (3, -1)$ and $f(1, 0) = (1, 2)$.

Exercise 3.4.4. Find $\operatorname{ran} f$, $\ker f$, $\operatorname{rank}(f)$, and $\dim \ker f$ in the following cases:

(1) $f : \mathbb{R}^2 \to \mathbb{R}^3$, $f(x, y) = (x - y, y, x + 2y)$.
(2) $f : \mathbb{R}^2 \to \mathbb{R}$, $f(x, y) = x - 2y$.
(3) $f : \mathbb{R}^3 \to \mathbb{R}^3$, $f(x, y, z) = (z, y, x)$.
(4) $f : \mathbb{R}^4 \to \mathbb{R}^4$, $f(x, y, z, t) = (x + y, x - z, z, t + x)$.

In the case f is bijective, determine its inverse.

Exercise 3.4.5. Let $f : \mathbb{R}_3[X] \to \mathbb{R}^3$ be defined by:

$$f(P) = (P(-1), P(0), P(1)).$$

(1) Show that f is linear.
(2) Is f surjective? Injective?

Exercise 3.4.6. Let $f : \mathbb{R}_2[X] \to \mathbb{R}_3[X]$ be defined by:
$$f[P(X)] = XP(X).$$
(1) Show that f is linear.
(2) Find a basis of $\ker f$ and a basis of $\operatorname{ran} f$. What is $\dim \ker f$ and $\dim \operatorname{ran} f$?
(3) Is f bijective?

Exercise 3.4.7. Let $E = \mathbb{R}[X]$ and let $f : E \to E$ be a map defined by:
$$f(P) = P - XP'.$$
(1) Check that f is linear.
(2) Find the kernel and the range of f.

Exercise 3.4.8. Let $f : \mathbb{R}_3[X] \to \mathbb{R}$ be a linear map defined by:
$$f(P) = \int_{-1}^{1} P(X)dX.$$
Is f one-to-one?

Exercise 3.4.9. Let $E = \mathcal{F}(\mathbb{R}, \mathbb{R})$ be the vector space of functions from $\mathbb{R}$ to $\mathbb{R}$. Let $T : E \to E$ be a map defined by:
$$T(f)(x) = f(x+1) - f(x).$$
(1) Prove that T is linear and find $\ker T$.
(2) Find $\ker(T^2)$, where $T^2 = T \circ T$.
(3) Determine $\ker T^n$ for each $n \in \mathbb{N}$.

Exercise 3.4.10. Let E be a vector space and let $f : E \to E$ be a projection. Inspired by arguments presented in Exercise 3.3.25 or else, show that $E = \operatorname{ran} f \oplus \ker f$.

Exercise 3.4.11. Find $\operatorname{ran} T$, where T is defined in Exercise 3.3.27.

Exercise 3.4.12. Let $f : E \to E$ be a linear map defined by:
$$f\left(\begin{pmatrix} a & b \\ c & d \end{pmatrix}\right) = \begin{pmatrix} d & -b \\ -c & a \end{pmatrix}.$$
(1) Show that the function f is injective, then explain why it is bijective.
(2) Find the expression of f^{-1}.

Exercise 3.4.13. Let $f : \mathbb{R}^3 \to \mathbb{R}^3$ be a linear map defined by:
$$f(x, y, z) = (x + 2y - z, -y, x + 7z).$$
(1) Find the matrix associated with f in the canonical basis of $\mathbb{R}^3$, denoted by $\{e_1, e_2, e_3\}$.
(2) Set
$$e_1' = e_1, \quad e_2' = e_1 + e_2, \quad e_3' = e_1 + e_2 + e_3.$$
Show that $\{e_1', e_2', e_3'\}$ is another basis of $\mathbb{R}^3$.
(3) Find the matrix associated with f in the basis $\{e_1', e_2', e_3'\}$.

Exercise 3.4.14. Let $E = M_2(\mathbb{R})$ and let:
$$M = \begin{pmatrix} 1 & 2 \\ 0 & 3 \end{pmatrix}.$$
We define three endomorphisms $f, g, h : E \to E$ by
$$f(A) = MA, \quad g(A) = AM, \quad h(A) = f(A) - g(A)$$
respectively.
(1) Determine the matrices associated with f, g, h in the canonical basis of E.
(2) Find $\ker h$, $\operatorname{ran} h$. What is $\dim \operatorname{ran} h$?

Exercise 3.4.15. Let E be a vector space of a finite dimension and let f and g be two endomorphisms of E. Prove that:
$$\operatorname{rank}(f + g) = \operatorname{rank}(f) + \operatorname{rank}(g) \Longleftrightarrow \begin{cases} \ker f + \ker g = \{0\}, \\ \operatorname{ran} f + \operatorname{ran} g = E. \end{cases}$$

Exercise 3.4.16. Let $E = C^1([0, 1], \mathbb{R})$ denote the vector space of real-valued functions on the interval $[0, 1]$ that are continuously differentiable on $[0, 1]$. Let $T : E \to \mathbb{R}$ be a map defined by:
$$T(f) = f(0) + 2f'(0) + 3f'(1).$$
(1) Show that T is a linear functional on E.
(2) Show that T is not injective.

Exercise 3.4.17. ([6]) Let $f, g \in L(E)$ obey $\operatorname{ran}(f) \cap \operatorname{ran}(g) = \{0\}$. Show that $\lambda f + \beta g$ is invertible if and only if $f + g$ is invertible, where λ and β are complex numbers.

Exercise 3.4.18. Let $f : E \to F$ be a mapping of complex vector spaces. *Call f anti-linear (or conjugate-linear) provided:*

- *For all $x, y \in E$: $f(x + y) = f(x) + f(y)$.*
- *For any complex scalar α and any $x \in E$: $f(\alpha x) = \overline{\alpha} f(x)$.*

(1) Show that $f : \mathbb{C} \to \mathbb{C}$, defined by $f(z) = \overline{z}$, is anti-linear map.

(2) Show that the composite of an anti-linear and a linear maps is anti-linear.

(3) Show that the composite of two anti-linear maps is *linear*.

CHAPTER 4

Determinants and Systems of Linear Equations

4.1. Course Summary

4.1.1. Determinants and their general properties.

We start with the definition of the determinant. The definition here is the so-called "Laplace expansion" (also known as the "cofactor expansion"), which expresses the determinant of a square matrix in terms of the determinants of its smaller submatrices (minors).

DEFINITION 4.1.1. Let $A \in M_n$. The determinant of $A := (a_{ij})$, where $1 \leq i, j \leq n$, is defined as:

$$\det(A) = \sum_{j=1}^{n} (-1)^{i+j} a_{ij} \det(A_{ij}),$$

where A_{ij} is the $(n-1) \times (n-1)$ matrix obtained by deleting the ith row and jth column of A (and $\det(A_{ij})$ is the determinant of the minor matrix A_{ij}).

In symbols, we can write:

$$\det(A) = \begin{vmatrix} a_{11} & a_{12} & \cdots & a_{1n} \\ a_{21} & a_{22} & \cdots & a_{2n} \\ \vdots & \vdots & \ddots & \vdots \\ a_{n1} & a_{n2} & \cdots & a_{nn} \end{vmatrix}.$$

EXAMPLE 4.1.1. Let

$$A = \begin{pmatrix} a_{11} & a_{12} & a_{13} \\ a_{21} & a_{22} & a_{23} \\ a_{31} & a_{32} & a_{33} \end{pmatrix}.$$

The Laplace expansion along the *first row* is:

$$\det(A) = a_{11} \det \begin{pmatrix} a_{22} & a_{23} \\ a_{32} & a_{33} \end{pmatrix} - a_{12} \det \begin{pmatrix} a_{21} & a_{23} \\ a_{31} & a_{33} \end{pmatrix} + a_{13} \det \begin{pmatrix} a_{21} & a_{22} \\ a_{31} & a_{32} \end{pmatrix}.$$

Below are several key properties of determinants.

THEOREM 4.1.1. *Let $A, B \in M_n(\mathbb{F})$ and let $\alpha \in \mathbb{F}$. Then*

(1) $\det A \in \mathbb{F}$.

(2) $\det(AB) = \det A \det B$.

(3) $\det(A) = \det(A^t)$.

(4) *If a row (or column) of matrix A is multiplied by α, the determinant of A is also multiplied by α. Thus,*

$$\det(\alpha A) = \alpha^n \det A, \ \forall \alpha \in \mathbb{F}.$$

(5) *Interchanging two rows (or columns) changes the sign of the determinant:*

$$\det(A') = -\det(A),$$

where A' is obtained by swapping two rows (or columns) of A.

(6) *Adding a scalar multiple of one row (or column) to another row (or column) does not change the determinant:*

$$\det(A) = \det(A'),$$

where A' is obtained by performing this row operation.

(7) *If f and g are linear transformations on a vector space E, and if A and B are the matrices representing f and g, then the determinant of their composition is given by $\det(AB)$.*

(8) *A is invertible if and only if $\det(A) \neq 0$. In such a case, $\det(A^{-1}) = 1/\det(A)$.*

(9) *The determinant of a triangular matrix, whether lower or upper—in particular, a diagonal matrix—is given by the product of its diagonal entries.*

Remark. We have repeatedly used the fact that a square matrix which is only right (or left) invertible is fully invertible. A proof of this appeared in Solution 3.3.34. Using the determinant, we can provide a simple alternative proof: If $AB = I$, then it follows that $\det A \cdot \det B = 1$, which implies that both $\det A \neq 0$ and $\det B \neq 0$. Therefore, both A and B are invertible.

Remark. Thanks to the formula $\det(A) = \det(A^t)$, we can alternatively expand the determinant along the jth column as follows:

$$\det(A) = \sum_{i=1}^{n}(-1)^{i+j}a_{ij}\det(A_{ij}).$$

Remark. Readers who will pursue a more advanced course on matrices with entries over rings will encounter specific matrices with entries in noncommutative rings where the matrix A is invertible, but its transpose A^t is not (hence $\det(A) \neq \det(A^t)$), see, e.g., [11]. I felt it was important to mention this for the sake of scientific rigor!

4.1.2. Adjugate formula for the inverse of a matrix.

Let A be an $n \times n$ invertible matrix. The cofactor C_{ij} of an entry a_{ij} is given by:

$$C_{ij} = (-1)^{i+j} \det(A_{ij}),$$

where A_{ij} is the $(n-1) \times (n-1)$ submatrix obtained by removing the ith row and jth column of A.

The inverse of A is given by the adjugate formula:

$$A^{-1} = \frac{1}{\det(A)} \operatorname{adj}(A),$$

where $\operatorname{adj}(A)$ is the adjugate (or adjoint) of A, which is the transpose of the cofactor matrix of A. In other words,

$$\operatorname{adj}(A) = \begin{pmatrix} C_{11} & C_{21} & \cdots & C_{n1} \\ C_{12} & C_{22} & \cdots & C_{n2} \\ \vdots & \vdots & \ddots & \vdots \\ C_{1n} & C_{2n} & \cdots & C_{nn} \end{pmatrix}.$$

EXAMPLES 4.1.1.

(1) For a 2×2 matrix $A = \begin{pmatrix} a & b \\ c & d \end{pmatrix}$, the adjugate formula for the inverse is:

$$A^{-1} = \frac{1}{\det(A)} \begin{pmatrix} d & -c \\ -b & a \end{pmatrix},$$

where $\det(A) = ad - bc \neq 0$.

(2) Let $A = \begin{pmatrix} a & b & c \\ d & e & f \\ g & h & i \end{pmatrix}$. Assume that $\det(A) \neq 0$, making A invertible. The cofactor matrix $\operatorname{Cof}(A)$ is:

$$\operatorname{Cof}(A) = \begin{pmatrix} \begin{vmatrix} e & f \\ h & i \end{vmatrix} & -\begin{vmatrix} d & f \\ g & i \end{vmatrix} & \begin{vmatrix} d & e \\ g & h \end{vmatrix} \\ -\begin{vmatrix} b & c \\ h & i \end{vmatrix} & \begin{vmatrix} a & c \\ g & i \end{vmatrix} & -\begin{vmatrix} a & b \\ g & h \end{vmatrix} \\ \begin{vmatrix} b & c \\ e & f \end{vmatrix} & -\begin{vmatrix} a & c \\ d & f \end{vmatrix} & \begin{vmatrix} a & b \\ d & e \end{vmatrix} \end{pmatrix}.$$

The adjugate of A is the matrix:

$$
\text{adj}(A) = \begin{pmatrix} \begin{vmatrix} e & f \\ h & i \end{vmatrix} & -\begin{vmatrix} b & c \\ h & i \end{vmatrix} & \begin{vmatrix} b & c \\ e & f \end{vmatrix} \\ -\begin{vmatrix} d & f \\ g & i \end{vmatrix} & \begin{vmatrix} a & c \\ g & i \end{vmatrix} & -\begin{vmatrix} a & c \\ d & f \end{vmatrix} \\ \begin{vmatrix} d & e \\ g & h \end{vmatrix} & -\begin{vmatrix} a & b \\ g & h \end{vmatrix} & \begin{vmatrix} a & b \\ d & e \end{vmatrix} \end{pmatrix}.
$$

Then one uses

$$
A^{-1} = \frac{1}{\det(A)} \, \text{adj}(A).
$$

Remark. Even though this method is liked by students for its theoretical elegance, the adjugate formula for finding matrix inverses is not practical for large matrices, because it requires a lot of time and memory, and can be unstable numerically. More efficient methods, such as Gaussian elimination, are typically used in real-world situations.

4.1.3. Rank of matrices.

DEFINITION 4.1.2. The rank of a matrix is the maximum number of linearly independent rows or columns.

In practice and using determinants, we can determine the rank of a matrix (square or rectangular) as follows:

(1) Compute the determinant of the largest square submatrices.
(2) The rank of the matrix is the size of the largest square submatrix with a nonzero determinant.

Remark. Here are more details on how to determine the rank of a matrix.

For square matrices, if the determinant is nonzero, the matrix is said to have full rank, which means its rank is equal to n (the size of the matrix). If the determinant is zero, we can reduce the size of the matrix by removing rows and columns until we find the largest nonzero determinant among its submatrices, if any exists.

For rectangular matrices (those that are not of full rank), consider all possible square submatrices (for example, 2×2, 3×3, etc.). Calculate the determinants of these submatrices. The rank of the matrix is determined by the size of the largest square submatrix that has a nonzero determinant.

4.1.4. Linear systems and Cramer's rule.

We have already solved systems of linear equations using methods familiar to readers at a lower level. In this section, we explore these systems in more detail, focusing on their structure and other solution methods to deepen their understanding of this key topic.

We start with the definition of a linear system. The general form of a linear system is:

$$\begin{cases} a_{11}x_1 + a_{12}x_2 + \cdots + a_{1n}x_n = b_1, \\ a_{21}x_1 + a_{22}x_2 + \cdots + a_{2n}x_n = b_2, \\ \vdots \\ a_{m1}x_1 + a_{m2}x_2 + \cdots + a_{mn}x_n = b_m. \end{cases}$$

In terms of matrices, the previous system may be expressed as $AX = B$, where

$$A = \begin{pmatrix} a_{11} & a_{12} & \cdots & a_{1n} \\ a_{21} & a_{22} & \cdots & a_{2n} \\ \vdots & \vdots & \ddots & \vdots \\ a_{m1} & a_{m2} & \cdots & a_{mn} \end{pmatrix}, \quad X = \begin{pmatrix} x_1 \\ x_2 \\ \vdots \\ x_n \end{pmatrix}, \quad \text{and } B = \begin{pmatrix} b_1 \\ b_2 \\ \vdots \\ b_m \end{pmatrix}.$$

Matrix A is called the coefficient matrix (of the system).

When $B = 0$, the system is said to be homogeneous and always has at least the trivial solution $X = 0$.

The final important topic of this chapter is the so-called Cramer's rule. Consider a system of n linear equations in n unknowns:

$$AX = B,$$

where A is an $n \times n$ invertible matrix, X is the column vector of unknowns, and B is the column vector of constants. Since A is invertible, the system $AX = B$ has a unique solution, given by

$$X = A^{-1}B \text{ (and not } BA^{-1}!).$$

If A^{-1} has already been computed, what follows becomes redundant. Otherwise, Cramer's rule provides a convenient method for determining $A^{-1}B$, though we do not provide the details here.

The unique solution $X = (x_1, x_2, \ldots, x_n)^t$ is given by:

$$x_i = \frac{\det(A_i)}{\det(A)} \quad \text{for } i = 1, 2, \ldots, n,$$

where A_i is the matrix obtained by replacing the ith *column* of A with the column vector B. This result is due to Cramer, and it will be referred to as Cramer's rule.

Remark. Cramer's rule is not efficient for large systems because it requires the calculation of multiple determinants. It is more suitable for small systems or theoretical applications.

4.2. True or False

Questions. Answer the following questions, providing justifications for your answers:

(1) Let A be a square matrix of order n and I be the identity matrix of the same order. Then

$$\det(I + A) = 1 + \det A.$$

(2) Let A be a square matrix of order n. Then

$$\det(A^2) = (\det A)^2.$$

(3) Let $A \in M_n$. Then

$$\det(-A) = -\det(A).$$

(4) Let $a \in \mathbb{R}$. We have

$$\begin{vmatrix} 2a & a & -a \\ a & 2a & a \\ a & 3a & a \end{vmatrix} = a \begin{vmatrix} 2 & 1 & -1 \\ 1 & 2 & 1 \\ 1 & 3 & 1 \end{vmatrix}.$$

(5) A determinant can be evaluated by expanding along any row or any column.

(6) A determinant that has a row or a column consisting entirely of zeros is always equal to zero.

(7) Let $A, B \in M_n(\mathbb{F})$. Then

$$\det(AB) = \det(BA) \Longleftrightarrow AB = BA.$$

(8) Let $A, B, C \in M_n(\mathbb{F})$. Then

$$\det(ABC) = \det(BAC).$$

(9) Two similar matrices have the same determinant.

(10) Let A be a nilpotent matrix. Then $\det A = 0$.

(11) The determinant map (from $M_n(\mathbb{R})$ to $\mathbb{R}$) is a polynomial.

(12) Let $A \in M_n$. There is always a $\lambda \in \mathbb{C}$ such that $A - \lambda I_n$ is not invertible.

(13) What is the Sarrus' rule for calculating determinants of 3×3 matrices?

(14) The Sarrus' rule can be extended to calculate determinants of higher orders matrices such as 4×4 ones.

(15) A system of linear equations always admits a unique solution.

(16) A system of n equations of n unknowns admits a unique solution if and only if the determinant of the coefficient matrix is nonzero.

(17) The solutions of the following system:

$$\begin{cases} x - y = \frac{1}{2}, \\ x + 2y = 5, \\ x + y = -4, \end{cases}$$

are $x = 2$ and $y = \frac{3}{2}$.

(18) A homogeneous system of the form $AX = 0$, where $\det A = 0$, has infinitely many solutions.

(19) A non-homogeneous system of the form $AX = B$, where $\det A = 0$, has infinitely many solutions.

(20) Let $A \in M_n(R)$, where R is a certain ring. Then

$$\det(A) \neq 0 \iff A \text{ is invertible.}$$

(21) Let $A \in M_n(\mathbb{R})$ be such that $\det A < 0$. Then A does not have any square roots in $M_n(\mathbb{R})$.

Answers.

(1) **False!** For example, take the matrix $A = \begin{pmatrix} 1 & 2 \\ 2 & 1 \end{pmatrix}$, then $A + I = \begin{pmatrix} 2 & 2 \\ 2 & 2 \end{pmatrix}$. So, $1 + \det A = 1 + (-3) = -2$, while $\det(A + I) = 0$.

   Remark. This example demonstrates that the mapping $\det : M_n(\mathbb{F}) \to \mathbb{F}$ is not linear.

(2) **True.** Since $\det(AB) = \det A \det B$, it ensues that

$$\det(A^2) = \det A \det A = (\det A)^2.$$

(3) **False, in general!** We know that $\det(-A) = (-1)^n \det A$, meaning that $\det(-A) = -\det(A)$ only when n (the matrix order) is odd. Therefore, a counterexample is only available when n is even. In fact, for even-order matrices, the only ones that satisfy $\det(-A) = -\det(A)$ are singular matrices, i.e., matrices where $\det(A) = 0$. In other words, the only way an even-order matrix can satisfy $\det(-A) = -\det(A)$ is if it is singular. Thus, any even-order non-singular matrix serves as a counterexample. To confirm this, consider the following

explicit counterexample: Let $A = \begin{pmatrix} 1 & 0 \\ 0 & 1 \end{pmatrix}$, which is a 2×2 identity matrix. Here, $\det A = 1$, yet

$$\det(-A) = \det \begin{pmatrix} -1 & 0 \\ 0 & -1 \end{pmatrix} = 1 \neq -1 = -\det(A).$$

(4) **False (unless $a = 0, 1, -1$)!** What is *true* is the following:

$$\begin{vmatrix} 2a & a & -a \\ a & 2a & a \\ a & 3a & a \end{vmatrix} = a^3 \begin{vmatrix} 2 & 1 & -1 \\ 1 & 2 & 1 \\ 1 & 3 & 1 \end{vmatrix}.$$

(5) **True.** Typically, we should select the row or column with the most zeros. Be cautious with the "+" and "−" signs that arise from $(-1)^{i+j}$, where i represents the row and j denotes the column.

If we want, for instance, to evaluate $\begin{vmatrix} 2 & 0 & 3 \\ 1 & 0 & 4 \\ 1 & -2 & 1 \end{vmatrix}$, we ex-

pand along the second column. The determinant then becomes (remembering that $(-1)^{3+2} = -1$)

$$-(-2) \begin{vmatrix} 2 & 3 \\ 1 & 4 \end{vmatrix} = +2(8 - 3) = 10.$$

We often attempt to create zeros in a row or column by using linear combinations of other rows (and/or columns). In some cases, we can even transform the matrix into an upper or lower triangular form. However, it isn't always necessary to do this. At times, calculating the determinant directly can be more straightforward and quicker than trying to create zeros.

(6) **True.** The reason is that we need only expand along that row or that column that contains only zeros. For instance, we have

$$\begin{vmatrix} -142 & 0 & 3589 & 1564 \\ 0 & 0 & 0 & 0 \\ 1 & -2 & 1 & 3654 \\ -15 & 56 & 89 & -71 \end{vmatrix} = 0.$$

(7) We always have the property that $\det(AB) = \det(BA)$ (even without commutativity of the matrices). This is because:

$$\det(AB) = \det A \cdot \det B = \det B \cdot \det A = \det(BA).$$

(by the commutativity of "·" over $\mathbb{F}$).

However, since $\det(AB) = \det(BA)$ holds for *any* A and B, we do not have necessarily $AB = BA$. As a counterexample, let

$$A = \begin{pmatrix} 1 & 0 \\ 1 & 0 \end{pmatrix} \text{ and } B = \begin{pmatrix} 0 & 1 \\ 0 & 1 \end{pmatrix},$$

and details are left to readers.

(8) **True.** Indeed,

$$\det(ABC) = \det A \det(BC) = \det A \det B \det C.$$

Since multiplication in $\mathbb{F}$ is commutative,

$$\det A \det B \det C = \det B \det A \det C.$$

Thus,

$$\det(ABC) = \det(BAC).$$

In another language, the determinant remains unchanged under cyclic permutations of the factors.

(9) **True.** To prove it, suppose matrix A is similar to another one B, i.e., for some invertible matrix P such that $B = P^{-1}AP$. Then

$$\det B = \det(P^{-1}) \cdot \det A \cdot \det P = (\det P)^{-1} \cdot \det A \cdot \det P$$

(because $\det(P^{-1}) = (\det P)^{-1}$). Hence

$$\det B = (\det P)^{-1} \cdot \det P \cdot \det A = \det A.$$

(10) **True.** By assumption, $A^k = 0$ for some $k \in \mathbb{N}$. Hence,

$$0 = \det(0) = \det(A^k) = (\det A)^k,$$

which gives $\det A = 0$.

(11) **True.** Such a determinant is a polynomial in n^2 variables, which is precisely the number of entries in each square matrix of size $n \times n$. In other words, these n^2 variables are

$$a_{11}, \ldots, a_{1n}, a_{21}, \ldots, a_{2n}, a_{n1}, \ldots, a_{nn}.$$

(12) **True.** The answer is simple once we recall the Fundamental Theorem of Algebra. Indeed, $\lambda \mapsto \det(A - \lambda I_n)$ is a (non-constant) polynomial in λ of degree n, and so it has at least one complex root. As simple as that.

(13) The Sarrus' rule is a straightforward method to compute the determinant of 3×3 matrices. Given a 3×3 matrix:

$$A = \begin{pmatrix} a & b & c \\ d & e & f \\ g & h & i \end{pmatrix},$$

the determinant, $\det(A)$, is calculated as:

$$\det(A) = aei + bfg + cdh - ceg - afh - bdi.$$

This expression is better memorized by considering the augmented determinant

$$\begin{vmatrix} a & b & c \\ d & e & f \\ g & h & i \end{vmatrix} \begin{matrix} a & b \\ d & e \\ g & h \end{matrix},$$

where we have taken the product of the first three diagonal-like elements, then we subtract the products of the off-diagonal-like elements.

Why does Sarrus' rule provide the exact value of the determinant? This can be clearly observed by expanding the determinant with respect to the first row as follows:

$$\begin{vmatrix} a & b & c \\ d & e & f \\ g & h & i \end{vmatrix} = a \begin{vmatrix} e & f \\ h & i \end{vmatrix} - b \begin{vmatrix} d & f \\ g & i \end{vmatrix} + c \begin{vmatrix} d & e \\ g & h \end{vmatrix}$$

$$= a(ei - fh) - b(di - fg) + c(dh - eg),$$

which is the above expression!

(14) **False!** Consider the matrix:

$$B = \begin{pmatrix} 1 & 2 & 0 & 1 \\ 0 & 1 & 3 & 2 \\ 4 & 0 & 1 & 3 \\ 1 & 2 & 3 & 1 \end{pmatrix}.$$

Readers may check that the value of this determinant is -45.

Now, to attempt to mimic Sarrus' rule for 4×4 matrices, we first augment the matrix by adding its first two columns to the right:

$$\begin{vmatrix} 1 & 2 & 0 & 1 \\ 0 & 1 & 3 & 2 \\ 4 & 0 & 1 & 3 \\ 1 & 2 & 3 & 1 \end{vmatrix} \begin{matrix} 1 & 2 \\ 0 & 1 \\ 4 & 0 \\ 1 & 2 \end{matrix}.$$

If a generalization of Sarrus' rule can be applied to 4×4 matrices, then the sum of the products of the first three diagonal-like elements, minus the product of the off-diagonal-like elements, denoted as $S(B)$, should equal the determinant of matrix B. However, since $S(B) = 15$, it does not match the previously calculated value of $\det(B)$.

Since the previous attempt was unsuccessful, we might consider exploring an alternative method by adding three columns (after all, in Sarrus' rule, we add *two* columns to a 3×3 matrix). So, consider

$$\begin{array}{cccc|ccc}
1 & 2 & 0 & 1 & 1 & 2 & 0 \\
0 & 1 & 3 & 2 & 0 & 1 & 3 \\
4 & 0 & 1 & 3 & 4 & 0 & 1 \\
1 & 2 & 3 & 1 & 1 & 2 & 3
\end{array}.$$

Readers may also observe that summing the products of the diagonal-like elements and then subtracting the products of the off-diagonal-like elements yields $15 \neq \det(B)$ (once again, the same value as $S(B)$ above, and is this just a coincidence?).

(15) **False!** Some systems admit a unique solution, others admit infinitely many, and others do not admit any solution! For example, the system:

$$\begin{cases} x - y = 1, \\ x + y = 1, \end{cases}$$

admits a unique solution $(1, 0)$, whereas the system

$$\begin{cases} x + y = 1, \\ 2x + 2y = 2, \end{cases}$$

admits infinitely many solutions because both equations reduce to the single equation $x + y = 1$. On the other hand, the system

$$\begin{cases} x + y = 0, \\ x + y = 1, \end{cases}$$

has no solution since the two equations contradict each other.

(16) **True.** This is a key result. For instance, it allows us to use Cramer's formula. Another practical consequence of this property is that sometimes a solution of a given system is apparent, and if the determinant is nonzero, then we immediately deduce that this is the only solution! For example, the system

$$\begin{cases} x - y + 2z = 2, \\ 2x + 3y - z = -1, \\ z - 4y = 1, \end{cases}$$

has $(0, 0, 1)$ as an apparent solution. The determinant of the coefficient matrix is given by

$$\Delta = \begin{vmatrix} 1 & -1 & 2 \\ 2 & 3 & -1 \\ 0 & -4 & 1 \end{vmatrix}$$

$$= 1 \begin{vmatrix} 3 & -1 \\ -4 & 1 \end{vmatrix} - 2 \begin{vmatrix} -1 & 2 \\ -4 & 1 \end{vmatrix} + 0 \begin{vmatrix} -1 & 2 \\ 3 & -1 \end{vmatrix},$$

so $\Delta = -15$. Since it is nonzero, we conclude that the solution $(0, 0, 1)$ is the *only* solution.

(17) **False!** The values $x = 2$ and $y = \frac{3}{2}$ satisfy the first and second equations, but not the third since $x + y = 2 + \frac{3}{2} = \frac{7}{2} \neq -4$. So, $(2, \frac{3}{2})$ is not a solution of the system!

(18) **True.** One way to see this is as follows: Since $\det(A) = 0$, causing A to be non-invertible, or equivalently, A to be non-injective. This means that $AX = 0$ for some nonzero X, and due to linearity, αX is a solution for each α (in $\mathbb{R}$, say). This way, we see that the number of solutions is infinite.

(19) **False!** Let us clarify this point. The equation $AX = B$ possesses a unique solution if and only if $\det A \neq 0$. Suppose that $B \neq 0$, as the case $B = 0$, has already been treated. When $\det A = 0$, the equation $AX = B$ may have an infinitude of solutions as it may have none! For example, consider

$$\begin{cases} x - y = 1, \\ x - y = 1, \end{cases} \text{ and } \begin{cases} x - y = 1, \\ x - y = 2. \end{cases}$$

Both systems have a zero determinant. The first system has an infinite number of solutions, represented by the family $(y+1, y)$ where $y \in \mathbb{R}$. In contrast, the second system has no solution at all.

(20) **False!** Only, the implication "$\Leftarrow$" is true. Before presenting a counterexample, readers should be aware that the equivalence "$\det(A) \neq 0$ if and only if A is invertible" is true if $A \in M_n(\mathbb{F})$, where $\mathbb{F}$ is a *field*. This information is essential because it helps us in our search for counterexamples. Consider the ring $\mathbb{Z}_4$, then define a matrix $A \in M_2(\mathbb{Z}_4)$ by $A = \begin{pmatrix} 1 & 0 \\ 0 & 2 \end{pmatrix}$. It is clear that $\det(A) = 2 \neq 0$. To see why A is not invertible, let $B = \begin{pmatrix} a & b \\ c & d \end{pmatrix}$ be such that $AB = \begin{pmatrix} 1 & 0 \\ 0 & 1 \end{pmatrix}$. This equation would yield (among others) $2 \cdot d = 1$, which is impossible, for

2 is not a unit (*a unit is an element that has a multiplicative inverse*). Thus, A is not invertible even if $\det(A) \neq 0$.

Before proving the reverse implication, let us provide another counterexample. Consider the matrix $A \in M_2(\mathbb{Z})$ defined by $A = \begin{pmatrix} 4 & 6 \\ 3 & 5 \end{pmatrix}$. Then, $\det(A) = 2$ (so, $\det(A) \neq 0$ and $\det(A) \neq \pm 1$). We can easily find that:

$$A^{-1} = \begin{pmatrix} \frac{5}{2} & -3 \\ -\frac{3}{2} & 2 \end{pmatrix}.$$

The entries of A^{-1} are not all integers because of the fractions $\frac{5}{2}$ and $-\frac{3}{2}$. Therefore, A^{-1} cannot be expressed with entries in $\mathbb{Z}$, even though $\det(A) \neq 0$. In other words, $A^{-1} \notin M_2(\mathbb{Z})$.

Remark. It is fairly easy to see that if $A \in M_n(\mathbb{Z})$ is invertible, then $A^{-1} \in M_n(\mathbb{Z})$ if and only if $\det A = \pm 1$.

Now, we show the other implication. Since defining the determinant of matrices over noncommutative rings is problematic, we assume that A has entries in a commutative ring (with identity). If A is invertible, then there exists a square matrix B of the same size such that $(BA =)\, I = AB$. Therefore, $\det(A)\det(B) = 1$, which implies that neither $\det(A)$ nor $\det(B)$ can be zero.

(21) **True.** Assume for contradiction that there exists a $B \in M_n(\mathbb{R})$ such that $A = B^2$. Then, we have

$$\det A = \det(B^2) = (\det B)^2.$$

Since $\det B \in \mathbb{R}$, it follows $(\det B)^2 \geq 0$. However, we are given that $\det A < 0$. This is a contradiction as the equation $\det A = (\det B)^2$ cannot be satisfied when $(\det B)^2 \geq 0$ and $\det A < 0$.

As an illustrative example, the matrix $A = \begin{pmatrix} 1 & 0 \\ 0 & -1 \end{pmatrix}$, whose determinant is (-1), does not have any square root $B \in M_2(\mathbb{R})$.

4.3. Exercises with Solutions

Exercise 4.3.1. Evaluate the following determinants:

$$\Delta_1 = \begin{vmatrix} 1 & -2 \\ -2 & -1 \end{vmatrix}, \quad \Delta_2 = \begin{vmatrix} 2 & 3 & 4 \\ 5 & 4 & 3 \\ 1 & 2 & 1 \end{vmatrix}.$$

Solution 4.3.1. We have:

$$\triangle_1 = \begin{vmatrix} 1 & -2 \\ -2 & -1 \end{vmatrix} = 1(-1) - (-2)(-2) = -1 - 4 = -5.$$

Also,

$$\triangle_2 = \begin{vmatrix} 2 & 3 & 4 \\ 5 & 4 & 3 \\ 1 & 2 & 1 \end{vmatrix} = +2 \begin{vmatrix} 4 & 3 \\ 2 & 1 \end{vmatrix} - 3 \begin{vmatrix} 5 & 3 \\ 1 & 1 \end{vmatrix} + 4 \begin{vmatrix} 5 & 4 \\ 1 & 2 \end{vmatrix}.$$

Thus,

$$\triangle_2 = 2(4 - 6) - 3(5 - 3) + 4(10 - 4) = -4 - 6 + 24 = 14.$$

Exercise 4.3.2. Compute the following determinants:

$$\triangle_1 = \begin{vmatrix} 2 & 3 & 4 \\ 5 & 6 & 7 \\ 8 & 9 & 1 \end{vmatrix}, \quad \triangle_2 = \begin{vmatrix} \frac{1}{2} & -1 & -\frac{1}{3} \\ \frac{3}{4} & \frac{1}{2} & -1 \\ 1 & -4 & 1 \end{vmatrix}.$$

Solution 4.3.2. To find $\triangle_1$, we replace the second row R_2 by $-2R_1 + R_2$, then we replace R_3 by $-3R_1 + R_3$. So,

$$\triangle_1 = \begin{vmatrix} 2 & 3 & 4 \\ 5 & 6 & 7 \\ 8 & 9 & 1 \end{vmatrix} = \begin{vmatrix} 2 & 3 & 4 \\ 1 & 0 & -1 \\ 8 & 9 & 1 \end{vmatrix} = \begin{vmatrix} 2 & 3 & 4 \\ 1 & 0 & -1 \\ 2 & 0 & -11 \end{vmatrix}.$$

We choose to expand with respect to the second column, i.e.,

$$\triangle_1 = \begin{vmatrix} 2 & 3 & 4 \\ 1 & 0 & -1 \\ 2 & 0 & -11 \end{vmatrix} = -3 \begin{vmatrix} 1 & -1 \\ 2 & -11 \end{vmatrix} + 0 - 0 = -3(-11 + 2) = 27.$$

The determinant $\triangle_2$ contains algebraic fractions, which may introduce errors during computation. To simplify, we multiply the determinant by $24 = 6 \times 4$, distributing 6 to the first row and 4 to the second row. Note that we will need to return to the original determinant $\triangle_2$ by dividing the final result by 24. After performing these adjustments, we find:

$$24\triangle_2 = \begin{vmatrix} \frac{6}{2} & -6 & -\frac{6}{3} \\ \frac{12}{4} & \frac{4}{2} & -4 \\ 1 & -4 & 1 \end{vmatrix} = \begin{vmatrix} 3 & -6 & -2 \\ 3 & 2 & -4 \\ 1 & -4 & 1 \end{vmatrix}.$$

After performing the operations $-R_1 + R_2 \to R_2$ and then $-3R_3 + R_1 \to R_1$, we obtain:

$$\begin{vmatrix} 3 & -6 & -2 \\ 3 & 2 & -4 \\ 1 & -4 & 1 \end{vmatrix} = \begin{vmatrix} 3 & -6 & -2 \\ 0 & 8 & -2 \\ 1 & -4 & 1 \end{vmatrix} = \begin{vmatrix} 0 & 6 & -5 \\ 0 & 8 & -2 \\ 1 & -4 & 1 \end{vmatrix} = +1 \begin{vmatrix} 6 & -5 \\ 8 & -2 \end{vmatrix}.$$

Thus,

$$24\Delta_2 = -12 + 40 = 28 \Longrightarrow \Delta_2 = \frac{28}{24} = \frac{7}{6}.$$

Exercise 4.3.3. Apply Sarrus' rule to calculate the following determinant:

$$\Delta = \begin{vmatrix} 2 & 3 & 4 \\ 5 & 4 & 3 \\ 1 & 2 & 1 \end{vmatrix}.$$

Solution 4.3.3. Write:

$$\begin{array}{ccc|cc} 2 & 3 & 4 & 2 & 3 \\ 5 & 4 & 3 & 5 & 4 \\ 1 & 2 & 1 & 1 & 2 \end{array}.$$

So

$$\Delta = 2 \times 4 \times 1 + 3 \times 3 \times 1 + 4 \times 5 \times 2 - 1 \times 4 \times 4 - 2 \times 3 \times 2 - 1 \times 5 \times 3,$$

that is, $\Delta = 14$.

Exercise 4.3.4. Prove that the following determinant is divisible by 13 without calculating its value:

$$\begin{vmatrix} 5 & 2 & 1 \\ 4 & 7 & 6 \\ 6 & 3 & 9 \end{vmatrix}.$$

Solution 4.3.4. We perform the following transformation: $100R_1 + 10R_2 + R_3 \to R_3$. Therefore, we arrive at:

$$\begin{vmatrix} 5 & 2 & 1 \\ 4 & 7 & 6 \\ 6 & 3 & 9 \end{vmatrix} = \begin{vmatrix} 5 & 2 & 1 \\ 4 & 7 & 6 \\ 546 & 273 & 169 \end{vmatrix} = 13 \begin{vmatrix} 5 & 2 & 1 \\ 4 & 7 & 6 \\ 42 & 21 & 13 \end{vmatrix},$$

which shows that the determinant is divisible by 13.

Exercise 4.3.5. Let $a, b, c \in \mathbb{R}$. Prove that

$$\Delta = \begin{vmatrix} 1+a & a & a \\ b & 1+b & b \\ c & c & 1+c \end{vmatrix} = 1 + a + b + c$$

without expanding the determinant.

Solution 4.3.5. We carry out the transformation $R_1 + R_2 + R_3 \rightarrow R_1$. After this transformation, we obtain:

$$\Delta = \begin{vmatrix} 1+a+b+c & 1+a+b+c & 1+a+b+c \\ b & 1+b & b \\ c & c & 1+c \end{vmatrix}.$$

So:

$$\Delta = (1+a+b+c) \underbrace{\begin{vmatrix} 1 & 1 & 1 \\ b & 1+b & b \\ c & c & 1+c \end{vmatrix}}_{\Delta'}.$$

To determine Δ', we will perform the following transformations:

(1) Replace R_2 with $R_2 - bR_1$.
(2) Replace R_3 with $R_3 - cR_1$.

Then,

$$\Delta' = \begin{vmatrix} 1 & 1 & 1 \\ b & 1+b & b \\ c & c & 1+c \end{vmatrix} = \begin{vmatrix} 1 & 1 & 1 \\ 0 & 1 & 0 \\ c & c & 1+c \end{vmatrix} = \begin{vmatrix} 1 & 1 & 1 \\ 0 & 1 & 0 \\ 0 & 0 & 1 \end{vmatrix}.$$

For an upper triangular matrix, the determinant is calculated as the product of the diagonal elements. Thus, we have $\Delta' = 1 \times 1 \times 1 = 1$. In conclusion:

$$\Delta = 1 + a + b + c.$$

Exercise 4.3.6. Suppose we are given the points $A(1,2)$, $B(4,5)$, and $C(6,3)$. Using the notion of determinant, compute the area of the parallelogram formed by the vectors derived from these points.

Solution 4.3.6. We only need three points to determine the area of a parallelogram in $\mathbb{R}^2$ since the fourth point is implicit; in case you're curious, the fourth point is given by $B + C - A$. Also, recall that the area of a parallelogram formed by two vectors $u = (u_1, u_2)$ and $v = (v_1, v_2)$ is given by

$$\Delta = |u_1 v_2 - u_2 v_1| = \left| \det \begin{pmatrix} u_1 & u_2 \\ v_1 & v_2 \end{pmatrix} \right|.$$

That being said, and as $\overrightarrow{AB}(3,3)$ and $\overrightarrow{AC}(5,1)$, we see that

$$\Delta = \left| \det \begin{pmatrix} 3 & 3 \\ 5 & 1 \end{pmatrix} \right| = |-12| = 12$$

square units.

Exercise 4.3.7. Calculate the volume of a parallelepiped S, denoted vol(S), in $\mathbb{R}^3$ determined by the three vectors $u = (1, 1, 0)$, $v = (1, 1, 1)$, and $w = (0, 2, 3)$.

Solution 4.3.7. We know that vol(S) = $|\det A|$, where

$$A = \begin{pmatrix} 1 & 1 & 0 \\ 1 & 1 & 1 \\ 0 & 2 & 3 \end{pmatrix}.$$

Then,

$$\det A = \begin{vmatrix} 1 & 1 & 0 \\ 1 & 1 & 1 \\ 0 & 2 & 3 \end{vmatrix} = -2,$$

so that:

$$\text{vol}(S) = |\det A| = |-2| = 2$$

cubic units.

Exercise 4.3.8. Show that the determinant map is surjective from $M_n(\mathbb{R})$ onto $\mathbb{R}$. Is it injective?

Solution 4.3.8. The determinant map is defined as follows: det : $M_n(\mathbb{R}) \to \mathbb{R}$, $A \mapsto \det A$. Surjectivity of det signifies that

$$\forall y \in \mathbb{R}, \ \exists A \in M_n(\mathbb{R}) : \ \det A = y.$$

Let y be real. We need to find a matrix A whose determinant equals y. One potential candidate is

$$A = \begin{pmatrix} y & 0 & \cdots & 0 \\ 0 & 1 & \ddots & \vdots \\ \vdots & \ddots & \ddots & 0 \\ 0 & \cdots & 0 & 1 \end{pmatrix}.$$

In this case, the determinant of the matrix is the product of its diagonal entries, which then simplifies to $y \times 1 \times \cdots \times 1 = y$.

The determinant function is not one-to-one. Let

$$A = \begin{pmatrix} 0 & 0 & \cdots & 0 \\ 0 & 1 & \ddots & \vdots \\ \vdots & \ddots & \ddots & 0 \\ 0 & \cdots & 0 & 0 \end{pmatrix}, \ B = \begin{pmatrix} 1 & 0 & \cdots & 0 \\ 0 & 0 & \ddots & \vdots \\ \vdots & \ddots & \ddots & 0 \\ 0 & \cdots & 0 & 0 \end{pmatrix}.$$

Although $A \neq B$, $\det A = \det B$ (=0).

Exercise 4.3.9. Let

$$A = \begin{pmatrix} a & b & c \\ d & e & f \\ g & h & i \end{pmatrix},$$

where all of its entries are real. Suppose that $\det(A) = -7$.

(1) Find

$$\det(3A), \ \det(A^{-1}), \ \det(2A^{-1}).$$

(2) Determine the determinant of each of the following matrices:

$$\begin{pmatrix} 2a & b & c \\ 2d & e & f \\ 2g & h & i \end{pmatrix}, \quad \begin{pmatrix} a & b & c \\ 3d & 3e & 3f \\ g & h & i \end{pmatrix}, \quad \begin{pmatrix} 3a & 3b & 3c \\ d & e & f \\ 3g & 3h & 3i \end{pmatrix}.$$

(3) Find

$$\det \begin{pmatrix} a+4d & b+4e & c+4f \\ d & e & f \\ g & h & i \end{pmatrix}.$$

(4) Find

$$\det \begin{pmatrix} a & g & d \\ b & h & e \\ c & i & f \end{pmatrix}.$$

Solution 4.3.9.

(1) Since A is of order 3, we have

$$\det(3A) = 3^3 \times \det A = 27 \times (-7) = -189.$$

To apply the formula for the inverse of a determinant, the matrix must be invertible. Since $\det A = -7 \neq 0$, it follows that A is invertible. Therefore,

$$\det(A^{-1}) = \frac{1}{\det A} = -\frac{1}{7}.$$

Finally,

$$\det(2A^{-1}) = 2^3 \times \det(A^{-1}) = -\frac{8}{7}.$$

(2) For the first matrix, we extract 2 only once; for the second, we extract 3 also once; and for the third, we extract 3 twice. Thus, we obtain:

$$\det \begin{pmatrix} 2a & b & c \\ 2d & e & f \\ 2g & h & i \end{pmatrix} = 2 \times \det A = -14,$$

$$\det \begin{pmatrix} a & b & c \\ 3d & 3e & 3f \\ g & h & i \end{pmatrix} = 3 \times \det A = -21,$$

and

$$\det \begin{pmatrix} 3a & 3b & 3c \\ d & e & f \\ 3g & 3h & 3i \end{pmatrix} = 3^2 \times \det A = -63.$$

(3) We have

$$\det \begin{pmatrix} a+4d & b+4e & c+4f \\ d & e & f \\ g & h & i \end{pmatrix} = \det A = -7.$$

(4) To find the determinant of the matrix, we start with matrix A and perform two steps. First, we take the transpose of A. Taking the transpose does not alter the value of the determinant. Second, we exchange the second and third columns. This exchange will result in the same determinant value but with a minus sign. Thus, we have:

$$\det \begin{pmatrix} a & g & d \\ b & h & e \\ c & i & f \end{pmatrix} = -(\det A) = -(-7) = +7.$$

Exercise 4.3.10. Let $A \in M_3(\mathbb{R})$ be such that $A^2 - 4A + 2I_3 = 0$, then set $B = 2A - 4I_3$. Find the possible values of $\det B$.

Solution 4.3.10. We can write $B = 2(A - 2I_3)$, and since B is a 3×3 matrix, it is seen $\det B = 2^3 \det(A - 2I_3)$. On the other hand,

$$(A - 2I_3)^2 = A^2 - 4A + 4I_3 = 2I_3,$$

whereby $\det(A - 2I_3)^2 = \det(2I_3) = 2^3 = 8$. Thus, $\det(A - 2I_3) = \pm 2\sqrt{2}$, and so $\det B = \pm 16\sqrt{2}$.

Exercise 4.3.11. Let A be a complex square matrix of order n, i.e.,

$$A = \begin{pmatrix} a_{11} & a_{12} & \cdots & a_{1n} \\ a_{21} & a_{22} & \cdots & a_{2n} \\ \cdots & \cdots & \cdots & \cdots \\ a_{n1} & a_{n2} & \cdots & a_{nn} \end{pmatrix},$$

where entries are possibly complex numbers. Set

$$\overline{A} = \begin{pmatrix} \overline{a_{11}} & \overline{a_{12}} & \cdots & \overline{a_{1n}} \\ \overline{a_{21}} & \overline{a_{22}} & \cdots & \overline{a_{2n}} \\ \cdots & \cdots & \cdots & \cdots \\ \overline{a_{n1}} & \overline{a_{n2}} & \cdots & \overline{a_{nn}} \end{pmatrix}.$$

(1) Show that $\det(\overline{A}) = \overline{\det A}$.
(2) Show that if A satisfies $A^t = \overline{A}$, then $\det(A) \in \mathbb{R}$.

Solution 4.3.11.

(1) By the definition of the determinant,

$$\det(\overline{A}) = \sum_{j=1}^{n} (-1)^{i+j} \overline{a_{ij}} \det(\overline{A_{ij}}).$$

The proof then follows by induction on n, which is left to interested readers.

(2) Since $A^t = \overline{A}$, it ensues that $\det(A^t) = \det(\overline{A})$. Additionally, $\det(A^t) = \det(A)$ and $\det(\overline{A}) = \overline{\det A}$. Thus, $\det(A) = \overline{\det A}$, which makes $\det(A)$ real, as needed.

Exercise 4.3.12. Find the rank of each of the following matrices:

$$A = \begin{pmatrix} 1 & 1 & 1 & 1 \\ 1 & 1 & -1 & -1 \\ 1 & -1 & 1 & -1 \\ 1 & -1 & -1 & -1 \end{pmatrix}, \quad B = \begin{pmatrix} 1 & 0 & -2 & -3 \\ -1 & -2 & 2 & -1 \\ 2 & -3 & 1 & -7 \\ 0 & 1 & 1 & 3 \end{pmatrix},$$

and

$$C = \begin{pmatrix} 1 & 3 & 5 & 1 \\ 2 & 2 & 6 & -2 \\ 1 & 2 & 4 & 0 \end{pmatrix}.$$

Solution 4.3.12. Since $\det A = -8 \neq 0$, we conclude that the rank of A equals its order; therefore, the rank of A is 4.

Let us calculate the determinant of B. We have:

$$B = \begin{vmatrix} 1 & 0 & -2 & -3 \\ -1 & -2 & 2 & -1 \\ 2 & -3 & 1 & -7 \\ 0 & 1 & 1 & 3 \end{vmatrix}$$

$$= (+1) \begin{vmatrix} -2 & 2 & -1 \\ -3 & 1 & -7 \\ 1 & 1 & 3 \end{vmatrix} + (-2) \begin{vmatrix} -1 & -2 & -1 \\ 2 & -3 & -7 \\ 0 & 1 & 3 \end{vmatrix}$$

$$+ -(-3) \begin{vmatrix} -1 & -2 & 2 \\ 2 & -3 & 1 \\ 0 & 1 & 1 \end{vmatrix}$$

$$= -12 - 2 \times 12 + 3 \times 12$$

$$= 0.$$

So, the rank of the given matrix is less than or equal to 3. Our calculations show that there is at least one non-null determinant of order 3, which indicates that the rank of the matrix B is 3.

The matrix C is not square, so we cannot calculate its determinant. However, we can evaluate the minors of order 3.

All minor determinants of order 3, which are:

$$\begin{vmatrix} 1 & 3 & 5 \\ 2 & 2 & 6 \\ 1 & 2 & 4 \end{vmatrix}, \quad \begin{vmatrix} 1 & 3 & 1 \\ 2 & 2 & -2 \\ 1 & 2 & 0 \end{vmatrix}, \quad \begin{vmatrix} 1 & 5 & 1 \\ 2 & 6 & -2 \\ 1 & 4 & 0 \end{vmatrix}, \quad \text{and} \quad \begin{vmatrix} 3 & 5 & 1 \\ 2 & 6 & -2 \\ 2 & 4 & 0 \end{vmatrix}$$

are zero. Thus, the rank is less than 3; that is, it is less than or equal to 2. Since, for example,

$$\begin{vmatrix} 2 & 2 \\ 1 & 2 \end{vmatrix} = 4 - 2 = 2 \neq 0,$$

we conclude that the rank of C is 2.

Exercise 4.3.13. ([9]) Let α be a real scalar, and let A_n be a square matrix of order n defined as follows:

$$A_n = \begin{pmatrix} \alpha & 1 & 1 & \cdots & 1 & 1 \\ \alpha & \alpha & 1 & \cdots & 1 & 1 \\ \vdots & \vdots & \vdots & & \vdots & \vdots \\ \alpha & \alpha & \alpha & \cdots & \alpha & 1 \\ \alpha & \alpha & \alpha & \cdots & \alpha & \alpha \end{pmatrix}, \quad D_n = \det A_n.$$

(1) Find D_n for $n = 2$ and $n = 3$.
(2) Prove that, for $n \geq 2$, $D_n = (\alpha - 1)D_{n-1}$, and find D_n for all $n \geq 1$.
(3) According to the values of α, determine the rank of A_n.

Solution 4.3.13.

(1) We have:

$$D_2 = \begin{vmatrix} \alpha & 1 \\ \alpha & \alpha \end{vmatrix} = \alpha^2 - \alpha = \alpha(\alpha - 1)$$

and

$$D_3 = \begin{vmatrix} \alpha & 1 & 1 \\ \alpha & \alpha & 1 \\ \alpha & \alpha & \alpha \end{vmatrix} = \begin{vmatrix} \alpha - 1 & 1 & 1 \\ 0 & \alpha & 1 \\ 0 & \alpha & \alpha \end{vmatrix} = (\alpha - 1) \begin{vmatrix} \alpha & 1 \\ \alpha & \alpha \end{vmatrix},$$

so

$$D_3 = (\alpha - 1)D_2 = \alpha(\alpha - 1)^2.$$

(2) Let $n \in \mathbb{N}$. We can write

$$D_n = \begin{vmatrix} \alpha & 1 & 1 & \cdots & 1 & 1 \\ \alpha & \alpha & 1 & \cdots & 1 & 1 \\ \vdots & \vdots & \vdots & & \vdots & \vdots \\ \alpha & \alpha & \alpha & \cdots & \alpha & 1 \\ \alpha & \alpha & \alpha & \cdots & \alpha & \alpha \end{vmatrix} = \begin{vmatrix} \alpha - 1 & 1 & 1 & \cdots & 1 & 1 \\ 0 & \alpha & 1 & \cdots & 1 & 1 \\ \vdots & \vdots & \vdots & & \vdots & \vdots \\ 0 & \alpha & \alpha & \cdots & \alpha & 1 \\ 0 & \alpha & \alpha & \cdots & \alpha & \alpha \end{vmatrix}$$

$$= (\alpha - 1) \underbrace{\begin{vmatrix} \alpha & 1 & 1 & \cdots & 1 & 1 \\ \alpha & \alpha & 1 & \cdots & 1 & 1 \\ \vdots & \vdots & \vdots & & \vdots & \vdots \\ \alpha & \alpha & \alpha & \cdots & \alpha & 1 \\ \alpha & \alpha & \alpha & \cdots & \alpha & \alpha \end{vmatrix}}_{\text{of order } (n-1) \times (n-1)}$$

$$= (\alpha - 1)D_{n-1}.$$

To calculate the general term D_n for all $n \geq 1$, we observe that (D_n) forms a geometric sequence with common ratio $\alpha - 1$, as it satisfies the recurrence relation $D_n = (\alpha - 1)D_{n-1}$. Given that the first term is D_2, the general formula can be derived as follows:

$$D_n = D_2(\alpha - 1)^{n-2}.$$

Substituting $D_2 = \alpha(\alpha - 1)$, we find:

$$D_n = \alpha(\alpha - 1)(\alpha - 1)^{n-2} = \alpha(\alpha - 1)^{n-1}.$$

Thus, the general term is given by:

$$D_n = \alpha(\alpha - 1)^{n-1},$$

valid for all $n \geq 1$. Note that this formula naturally includes $D_1 = \alpha$, as substituting $n = 1$ yields $\alpha(\alpha - 1)^0 = \alpha$.

(3) We will discuss this according to the values of α.

  (a) If $\alpha = 0$, then $D_n = 0$. So, rank $A_n \leq n - 1$. But, the determinant

$$\begin{vmatrix} 1 & 1 & \cdots & 1 & 1 \\ 0 & 1 & \cdots & 1 & 1 \\ \vdots & \vdots & & \vdots & \vdots \\ 0 & 0 & \cdots & 0 & 1 \end{vmatrix},$$

  of order $n - 1$, and extracted from A_n, is equal to 1, since it is the determinant of an upper triangular matrix. Therefore, the rank of A_n (when $\alpha = 0$) is $n - 1$.

  (b) If $\alpha = 1$, then it is evident that all determinants of order 2 or higher are zero. Since the matrix is not the zero matrix, we can conclude that the rank of A_n is 1.

  (c) If $\alpha \neq 0$ *and* $\alpha \neq 1$, then $D_n = \det A_n \neq 0$. In this case the rank is n.

Exercise 4.3.14. Find, using determinants, the inverse of the following matrices:

$$A = \begin{pmatrix} 2 & 1 \\ 2 & 3 \end{pmatrix} \quad \text{and} \quad B = \begin{pmatrix} -7 & 3 & 3 \\ -6 & 2 & 3 \\ -12 & 6 & 5 \end{pmatrix}.$$

Solution 4.3.14. We have: $\det A = 4 \neq 0$, so: A is invertible. The inverse of A is given by:

$$A^{-1} = \frac{1}{\det A} \begin{pmatrix} +3 & -2 \\ -1 & +2 \end{pmatrix}^t = \frac{1}{4} \begin{pmatrix} 3 & -1 \\ -2 & 2 \end{pmatrix}.$$

Since $\det B = 2 \neq 0$, B is invertible. Its inverse is given by:

$$B^{-1} = \frac{1}{\det B} \begin{pmatrix} +\begin{vmatrix} 2 & 3 \\ 6 & 5 \end{vmatrix} & -\begin{vmatrix} -6 & 3 \\ -12 & 5 \end{vmatrix} & +\begin{vmatrix} -6 & 2 \\ -12 & 6 \end{vmatrix} \\ -\begin{vmatrix} 3 & 3 \\ 6 & 5 \end{vmatrix} & +\begin{vmatrix} -7 & 3 \\ -12 & 5 \end{vmatrix} & -\begin{vmatrix} -7 & 3 \\ -12 & 6 \end{vmatrix} \\ +\begin{vmatrix} 3 & 3 \\ 2 & 3 \end{vmatrix} & -\begin{vmatrix} -7 & 3 \\ -6 & 3 \end{vmatrix} & +\begin{vmatrix} -7 & 3 \\ -6 & 2 \end{vmatrix} \end{pmatrix}^t$$

$$= \frac{1}{2} \begin{pmatrix} -8 & -6 & -12 \\ +3 & 1 & +6 \\ 3 & +3 & 4 \end{pmatrix}^t$$

$$= \frac{1}{2} \begin{pmatrix} -8 & 3 & 3 \\ -6 & 1 & 3 \\ -12 & 6 & 4 \end{pmatrix}.$$

Exercise 4.3.15. Show that the matrix

$$A = \begin{pmatrix} \cos\alpha & \sin\alpha & 0 \\ -\sin\alpha & \cos\alpha & 0 \\ 0 & 0 & 1 \end{pmatrix}$$

is invertible for all values of α. Find A^{-1}.

Solution 4.3.15. We show that the determinant of A is non-null whichever value of α. We expand with respect to the last row; we have for all $\alpha \in \mathbb{R}$:

$$\begin{vmatrix} \cos\alpha & \sin\alpha & 0 \\ -\sin\alpha & \cos\alpha & 0 \\ 0 & 0 & 1 \end{vmatrix} = +1 \begin{vmatrix} \cos\alpha & \sin\alpha \\ -\sin\alpha & \cos\alpha \end{vmatrix} = \cos^2\alpha + \sin^2\alpha = 1,$$

i.e., $\det A = 1 \neq 0$ (for all values of α). Hence, A is invertible. As in Exercise 4.3.14, we have

$$A^{-1} = \begin{pmatrix} \cos\alpha & \sin\alpha & 0 \\ -\sin\alpha & \cos\alpha & 0 \\ 0 & 0 & 1 \end{pmatrix}^t = \begin{pmatrix} \cos\alpha & -\sin\alpha & 0 \\ \sin\alpha & \cos\alpha & 0 \\ 0 & 0 & 1 \end{pmatrix}.$$

Exercise 4.3.16. Consider the following matrix:

$$A_a = \begin{pmatrix} 1 & 3 & a \\ 2 & -1 & 1 \\ -1 & 1 & 0 \end{pmatrix},$$

where a is a real scalar. Determine the linear map associated with A_a and show that it is bijective if and only if $a \neq 4$.

Solution 4.3.16. The associated linear transformation is defined from $\mathbb{R}^3$ into $\mathbb{R}^3$ by:

$$f_a(x, y, z) = (x + 3y + az, 2x - y + z, -x + y).$$

We also know that f_a is bijective if and only if $\det A_a$ is non-null. So, let's find the value of $\det A_a$ by expanding it with respect to the last row. We find:

$$\begin{vmatrix} 1 & 3 & a \\ 2 & -1 & 1 \\ -1 & 1 & 0 \end{vmatrix} = (-1) \begin{vmatrix} 3 & a \\ -1 & 1 \end{vmatrix} - 1 \begin{vmatrix} 1 & a \\ 2 & 1 \end{vmatrix} + 0 \begin{vmatrix} 1 & 3 \\ 2 & -1 \end{vmatrix}$$

$$= (-1)(3 + a) - (1 - 2a) = a - 4.$$

Thus, f_a is bijective if and only if $a \neq 4$.

Exercise 4.3.17. Determine the value(s) of a for which the following system of equations:

$$\begin{cases} x + y = 2, \\ 2x + 2y = a, \end{cases}$$

has:

(1) infinitely many solutions,
(2) no solutions,
(3) a unique solution.

Solution 4.3.17.

(1) The given system admits infinitely many solutions if and only if $a = 4$. In this case, the system reduces to the single equation $x + y = 2$. The set of solutions is given by:

$$\{(x, 2 - x) : x \in \mathbb{R}\}.$$

(2) The given system does not admit any solutions if $a \neq 4$. This is because the second equation $2x + 2y = a$ is inconsistent with $x + y = 2$ when $a \neq 4$.

(3) The given system cannot admit a unique solution for any value of a. This is due to the fact that the two equations are dependent (the second equation is a scalar multiple of the first), making the system underdetermined.

Exercise 4.3.18. Consider the system

$$\begin{cases} -x + y + z = 1, \\ x - y + z = 2, \\ x + y - z = -1. \end{cases}$$

(1) Write the previous system in matrix form, i.e., $AX = B$.
(2) Find the inverse of A.
(3) Deduce the solution of the given system.

Solution 4.3.18.

(1) We can write:

$$AX = B, \text{ where } A = \begin{pmatrix} -1 & 1 & 1 \\ 1 & -1 & 1 \\ 1 & 1 & -1 \end{pmatrix}, \ X = \begin{pmatrix} x \\ y \\ z \end{pmatrix},$$

$$\text{and } B = \begin{pmatrix} 1 \\ 2 \\ -1 \end{pmatrix}.$$

(2) We easily check that A is invertible and find that:

$$A^{-1} = \frac{1}{2} \begin{pmatrix} 0 & 1 & 1 \\ 1 & 0 & 1 \\ 1 & 1 & 0 \end{pmatrix}$$

(the computation details are left to the reader).

(3) Since A is invertible, the given system admits a unique solution. To find it, we proceed as follows:

$$AX = B \Longleftrightarrow X = A^{-1}B$$

$$\Longleftrightarrow \begin{pmatrix} x \\ y \\ z \end{pmatrix} = \frac{1}{2} \begin{pmatrix} 0 & 1 & 1 \\ 1 & 0 & 1 \\ 1 & 1 & 0 \end{pmatrix} \begin{pmatrix} 1 \\ 2 \\ -1 \end{pmatrix}$$

$$\Longleftrightarrow \begin{pmatrix} x \\ y \\ z \end{pmatrix} = \begin{pmatrix} \frac{1}{2} \\ 0 \\ \frac{3}{2} \end{pmatrix}.$$

Exercise 4.3.19. In $\mathbb{R}^3$, find the intersection of the planes defined by the following equations: $x - y + 3z + 2 = 0$, $x + z = 0$, and $x + 4y - z - 2 = 0$.

Solution 4.3.19. To find the intersection of the three planes in $\mathbb{R}^3$, we solve the given system of equations:

$$\begin{cases} x - y + 3z + 2 = 0, \\ x + z = 0, \\ x + 4y - z - 2 = 0. \end{cases}$$

We have:

$$\begin{cases} x - y + 3z + 2 = 0, \\ x + z = 0, \\ x + 4y - z - 2 = 0, \end{cases} \Longleftrightarrow \begin{cases} -z - y + 3z + 2 = 0, \\ x = -z, \\ -z + 4y - z - 2 = 0, \end{cases}$$

$$\Longleftrightarrow \begin{cases} -y + 2z + 2 = 0, \\ x = -z, \\ 4y - 2z - 2 = 0, \end{cases}$$

$$\iff \begin{cases} -y + 2z + 2 = 0, \\ x = -z, \\ 3y = 0, \end{cases}$$

$$\iff \begin{cases} 2z + 2 = 0, \\ x = -z, \\ y = 0, \end{cases}$$

$$\iff \begin{cases} x = 1, \\ y = 0, \\ z = -1. \end{cases}$$

Thus, these planes intersect at only one point, given by $(1, 0, -1)$.

Exercise 4.3.20. Using Cramer's rule, solve the system:
$$\begin{cases} 3x + 2y + z = 1, \\ x + 3y + 2z = 0, \\ 2x + y + 3z = -1. \end{cases}$$

Solution 4.3.20. The coefficient matrix is:

$$A = \begin{pmatrix} 3 & 2 & 1 \\ 1 & 3 & 2 \\ 2 & 1 & 3 \end{pmatrix}.$$

Its determinant is $\det A = 18$ (verify this calculation!). Since the determinant is nonzero, the system admits a unique solution. Using Cramer's rule, we know that

$$x = \frac{\det(A_1)}{\det(A)}, \quad y = \frac{\det(A_2)}{\det(A)}, \quad z = \frac{\det(A_3)}{\det(A)},$$

that is,

$$x = \frac{\begin{vmatrix} 1 & 2 & 1 \\ 0 & 3 & 2 \\ -1 & 1 & 3 \end{vmatrix}}{\det A} = \frac{6}{18} = \frac{1}{3},$$

$$y = \frac{\begin{vmatrix} 3 & 1 & 1 \\ 1 & 0 & 2 \\ 2 & -1 & 3 \end{vmatrix}}{\det A} = \frac{6}{18} = \frac{1}{3},$$

and

$$z = \frac{\begin{vmatrix} 3 & 2 & 1 \\ 1 & 3 & 0 \\ 2 & 1 & -1 \end{vmatrix}}{\det A} = \frac{-12}{18} = -\frac{2}{3}.$$

Exercise 4.3.21. ([9]) Let $a, b \in \mathbb{R}$. Solve the following system:

$$\begin{cases} x + ay = 2a \cdots (I), \\ bx + 2y + az = 4 - a^3 \cdots (II), \\ x + ay + bz = 0 \cdots (III). \end{cases}$$

Solution 4.3.21. Let us first evaluate the determinant of the coefficient matrix. It is given by:

$$\begin{vmatrix} 1 & a & 0 \\ b & 2 & a \\ 1 & a & b \end{vmatrix} = 1 \times \begin{vmatrix} 2 & a \\ a & b \end{vmatrix} - a \times \begin{vmatrix} b & a \\ 1 & b \end{vmatrix} = 2b - a^2 + a^2 - ab^2 = b(2 - ab).$$

The system admits a unique solution if and only if $b(2 - ab) \neq 0$, that is, if and only if $b \neq 0$ *and* $ab \neq 2$.

Using Cramer's rule, we can determine the values of x, y, and z. However, to save time, we can find z directly since the coefficients of x and y are the same in the first and third equations.

By subtracting Equation (I) from Equation (III), we obtain:

$$bz = -2a \text{ or } z = \frac{-2a}{b}.$$

We have:

$$x = \frac{\begin{vmatrix} 2a & a & 0 \\ 4 - a^3 & 2 & a \\ 0 & a & b \end{vmatrix}}{b(2 - ab)} = \frac{2a(2b - a^2) - a[(4 - a^3)b - 0]}{b(2 - ab)} = -\frac{a^3}{b}$$

and

$$y = \frac{\begin{vmatrix} 1 & 2a & 0 \\ b & 4 - a^3 & a \\ 1 & 0 & b \end{vmatrix}}{b(2 - ab)}$$

$$= \frac{1 \times [(4 - a^3)b - 0] - 2a(b^2 - a)}{b(2 - ab)} = \frac{a^2(2 - ab) + 2b(2 - ab)}{b(2 - ab)}$$

$$= \frac{a^2 + 2b}{b}.$$

So, when $b \neq 0$ *and* $ab \neq 2$, then the unique solution is given by:

$$(x, y, z) = \left(-\frac{a^3}{b}, \frac{a^2 + 2b}{b}, \frac{-2a}{b} \right).$$

Now, let us analyze the special cases $b = 0$ and $ab = 2$:

(1) **Case 1:** $b = 0$. Substituting $b = 0$ into the initial system (and not into Cramer's formula!), the system becomes:

$$\begin{cases} x + ay = 2a, \\ 2y + az = 4 - a^3, \\ x + ay = 0. \end{cases}$$

The left-hand sides of the first and third equations are identical, so we observe:

(a) If $a = 0$, then $x = 0$, $y = 2$, and $z \in \mathbb{R}$. Hence, the system admits infinitely many solutions, given by $\{(0, 2, z) : z \in \mathbb{R}\}$.

(b) If $a \neq 0$, then the system is inconsistent and admits no solutions.

(2) **Case 2:** $ab = 2$. Since $ab = 2$, it follows that $b \neq 0$ and $a \neq 0$. Substituting $b = \frac{2}{a}$ into the system, we have:

$$\begin{cases} x + ay = 2a, \\ \frac{2}{a}x + 2y + az = 4 - a^3, \\ x + ay + \frac{2}{a}z = 0. \end{cases}$$

From earlier calculations (as shown before applying Cramer's rule), we know $z = -a^2 = -\frac{2a}{b}$. Substituting this value of z into the system, we obtain:

$$\begin{cases} x + ay = 2a, \\ \frac{2}{a}x + 2y - a \cdot a^2 = 4 - a^3, \\ x + ay - \frac{2}{a} \cdot a^2 = 0, \end{cases}$$

which simplifies to:

$$\begin{cases} \frac{1}{a}x + y = 2, \\ x + ay = 2a, \end{cases}$$

and further reduces to a single equation $x + ay = 2a$. Solving for x, we get $x = 2a - ay$. Thus, the system admits infinitely many solutions, given by:

$$(x, y, z) = \left(2a - ay, y, -\frac{2a}{b}\right), \quad y \in \mathbb{R}.$$

Exercise 4.3.22. Solve the following systems in E:

(1) $E = \mathbb{Z}_5$ and

$$\begin{cases} 2x + y = 1, \\ x + 2y = 0. \end{cases}$$

(2) $E = \mathbb{Z}$ and

$$\begin{cases} 2x + y = 3, \\ x + 2y = 2. \end{cases}$$

(3) $E = \mathbb{Z}_4$ and

$$\begin{cases} x + 3y = 0, \\ x + y = 2. \end{cases}$$

Solution 4.3.22.

(1) Since $\mathbb{Z}_5$ is a field, the given system has a unique solution if and only if $\det A \neq 0$, where A is the matrix of the given system. Before carrying on, notice that all calculations are performed modulo 5. The coefficient matrix is $A = \begin{pmatrix} 2 & 1 \\ 1 & 2 \end{pmatrix}$. Its determinant is 3, which is invertible in $\mathbb{Z}_5$. Thus, the given system possesses a unique solution (x, y). Let us use Cramer's formula to find this solution. Notice that we will be "dividing" by 3, that is, we will be multiplying by the inverse of 3 in $\mathbb{Z}_5$. The (unique) inverse of 3, denoted by a, satisfies $3a = 1$. In $\mathbb{Z}_5$, only $a = 2$ satisfies the previous equation. So, dividing by 3 means that we are multiplying by 2. Keeping this in mind, we now compute the solutions as follows:

$$x = 2 \begin{vmatrix} 1 & 1 \\ 0 & 2 \end{vmatrix} = 4 \text{ and } y = 2 \begin{vmatrix} 2 & 1 \\ 1 & 0 \end{vmatrix} = 2 \times (-1) = 2 \times 4 = 3.$$

Thus, the solution to the system is $(x, y) = (4, 3)$.

(2) The given system has, over $\mathbb{Q}$, the unique solution $(x, y) = (4/3, 1/3)$. Thus, the given system has no solution over $\mathbb{Z}$. Observe in the end that the determinant of the coefficient matrix is nonzero.

(3) It can be checked that $(1, 1)$ and $(3, 3)$ are *two* solutions of the given system. However, the determinant of the coefficient matrix is $-2 = 2$ (over $\mathbb{Z}_4$), which is invertible. In this regard, there is no contradiction, as $\mathbb{Z}_4$ is not a field!

4.4. Exercises without Solutions

Exercise 4.4.1. ([**7**]) Calculate the following determinants:

$$\triangle_1 = \begin{vmatrix} 56372 & 56472 \\ 29413 & 29513 \end{vmatrix}, \quad \triangle_2 = \begin{vmatrix} 235 & 612 & 512 \\ 1035 & 1112 & 1012 \\ -365 & 812 & 712 \end{vmatrix}.$$

Exercise 4.4.2. Let

$$A = \begin{pmatrix} a & b & c \\ d & e & f \\ g & h & i \end{pmatrix},$$

where all entries are real scalars. Assume that $\det(A) = -6$. Evaluate the determinants of the following matrices:

$$\begin{pmatrix} d & e & f \\ g & h & i \\ a & b & c \end{pmatrix}, \quad \begin{pmatrix} 3a & 3b & 3c \\ -d & -e & -f \\ 4g & 4h & 4i \end{pmatrix},$$

and

$$\begin{pmatrix} -3a & -3b & -3c \\ d & e & f \\ g-4d & h-4e & i-4f \end{pmatrix}.$$

Exercise 4.4.3. Prove, without calculation, that the determinant:

$$\begin{vmatrix} 0 & 1 & 6 & 9 \\ 2 & -7 & 0 & 0 \\ 1 & 3 & 1 & 3 \\ 5 & 2 & 6 & 5 \end{vmatrix}$$

is divisible by 13.

Exercise 4.4.4. Let $a, b, c \in \mathbb{R}$. Prove that

$$\begin{vmatrix} b+c & c+a & b+a \\ a & b & c \\ 1 & 1 & 1 \end{vmatrix} = 0$$

without expanding it.

Exercise 4.4.5. Let $x \in \mathbb{R}$. Solve the equation:

$$\begin{vmatrix} x & -1 \\ 3 & 1-x \end{vmatrix} = \begin{vmatrix} 1 & 0 & -3 \\ 2 & x & -6 \\ 1 & 3 & x-5 \end{vmatrix}.$$

Exercise 4.4.6. (Vandermonde's determinant) Let $n \in \mathbb{N}$. Let $x_1, x_2, \ldots, x_n \in \mathbb{R}$. Show that the following determinant (said of Vandermonde)

$$\begin{vmatrix} 1 & 1 & \cdots & \cdots & 1 \\ x_1 & x_2 & \cdots & \cdots & x_n \\ x_1^2 & x_2^2 & \cdots & \cdots & x_n^2 \\ \vdots & \vdots & & & \vdots \\ x_1^{n-1} & x_2^{n-1} & \cdots & \cdots & x_n^{n-1} \end{vmatrix}$$

is equal to: $\prod_{1 \leq i < j \leq n}(x_i - x_j)$.

Exercise 4.4.7. Let x be a real scalar. Find the following *tridiagonal* determinant:

$$\begin{vmatrix} 1 + x^2 & -x & 0 & \cdots & & 0 \\ -x & 1 + x^2 & -x & \ddots & & \vdots \\ 0 & \ddots & \ddots & \ddots & & 0 \\ \vdots & \ddots & & -x & 1 + x^2 & -x \\ 0 & \cdots & & 0 & -x & 1 + x^2 \end{vmatrix}.$$

Exercise 4.4.8. Find the inverse of the following matrix:

$$A = \begin{pmatrix} 1 & 1 & 1 \\ 1 & 1 & 0 \\ -1 & 0 & 1 \end{pmatrix}.$$

Exercise 4.4.9. Let $a \in \mathbb{R}$. Discuss, according to the values of a, the rank of the following matrix:

$$A = \begin{pmatrix} 2 & a & a & a \\ a & 2 & a & a \\ a & a & 2 & a \\ a & a & a & 2 \end{pmatrix}.$$

Exercise 4.4.10. Let f_1, f_2, f_3, and f_4 be differentiable functions. Set

$$A = \begin{vmatrix} f_1(x) & f_2(x) \\ g_1(x) & g_2(x) \end{vmatrix}.$$

Show that

$$\frac{dA}{dx} = \begin{vmatrix} f_1'(x) & f_2'(x) \\ g_1(x) & g_2(x) \end{vmatrix} + \begin{vmatrix} f_1(x) & f_2(x) \\ g_1'(x) & g_2'(x) \end{vmatrix}.$$

Exercise 4.4.11. *Recall that if $\{f_1(x), f_2(x), f_3(x)\}$ is a set of twice differentiable functions on $\mathbb{R}$, then the* **Wronskian** *is given by:*

$$W(x) = \begin{vmatrix} f_1(x) & f_2(x) & f_3(x) \\ f_1'(x) & f_2'(x) & f_3'(x) \\ f_1''(x) & f_2''(x) & f_3''(x) \end{vmatrix}.$$

We also recall a fundamental result: If the functions of $\{f_1(x), f_2(x), f_3(x)\}$ are twice differentiable on $\mathbb{R}$, and if their Wronskian *is not identically null*, then the vectors are linearly independent.

Remark. The same result holds for a set of n vectors.

Apply the previous result to see whether the following sets are free:

(1) $\{x, \sin x\}$.
(2) $\{1, e^x, e^{2x}\}$.

Exercise 4.4.12. Apply the method of Exercise 4.4.11 to the sets of Exercise 2.3.22.

Exercise 4.4.13. In $\mathbb{R}^2$, find the intersection points of the two lines given by the equations: $x - y + 2 = 0$ and $x + y - 1 = 0$.

Exercise 4.4.14. Let $a \in \mathbb{R}$. Solve, according to the values of a, the system:
$$\begin{cases} x + y + (1 - a)z = a + 2, \\ (1 + a)x - y + 2z = 0, \\ 2x - ay + 3z = a + 2. \end{cases}$$

Exercise 4.4.15. Let $a \in \mathbb{C}$. Solve, according to the values of a, the system:
$$\begin{cases} x - ay + a^2 z = a, \\ ax - a^2 y + az = 1, \\ ax + y - a^3 z = 1. \end{cases}$$

Exercise 4.4.16. Solve the system:
$$\begin{cases} \sin x + 2\cos y + 3\tan z = 0, \\ 2\sin x + 5\cos y + 3\tan z = 0, \\ -\sin x - 5\cos y + 5\tan z = 0, \end{cases}$$
where $0 \le x, y \le 2\pi$, and $0 \le z < \frac{\pi}{2}$.

Exercise 4.4.17. Solve the system:
$$\begin{cases} x + 2\ln y + \sqrt{z} = 1, \\ x + \ln y - \sqrt{z} = 0, \\ -x - \ln y + 2\sqrt{z} = 1. \end{cases}$$

CHAPTER 5

Special Matrices

In this chapter, we introduce certain classes of matrices in an elementary setting, aiming to introduce readers to these concepts at a fundamental level, thus preparing them to delve deeper into them in subsequent courses. In particular, we only define these classes for real matrices.

5.1. Course Summary

DEFINITION 5.1.1. Let A be a square matrix. Say that A is:

(1) symmetric $A^t = A$;
(2) skew-symmetric if $A^t = -A$;
(3) normal when $A^t A = AA^t$;
(4) orthogonal $A^t A = I$.

Remark. If A is orthogonal, $A^t = A^{-1}$. Also, symmetric and skew-symmetric matrices are normal (see Exercise 5.3.15 for a proof).

Remark. While our discussion primarily focuses on matrices with real entries, readers should note that the above concepts naturally extend to the complex case, where matrices may have complex entries. To address this, we introduce the notion of the adjoint, which coincides with the transpose when all entries are real. The adjoint of a matrix A is defined as $A^* = \overline{A}^t$, where $\overline{A}$ was introduced in Exercise 4.3.11. We then say that A is Hermitian provided $A^* = A$ and A is normal if $AA^* = A^*A$.

Remark. Although real normal (or symmetric) matrices are essential in many applications, the complex case offers richer structures and tools, especially in fields involving complex vector spaces or other physical phenomena.

5.2. True or False

Questions. Answer the following questions, providing justifications for your answers:

(1) A matrix cannot be simultaneously both symmetric and skew-symmetric.

(2) A skew-symmetric matrix can have nonzero elements on its main diagonal.

(3) Let $A, B \in M_n(\mathbb{R})$ be both symmetric such that $AB = 0$. Then $BA = 0$.

(4) The inverse of a symmetric matrix is itself symmetric.

(5) Let A be a non-necessarily square matrix. Then $A^t A$ and AA^t are both symmetric.

(6) Let $A \in M_n(\mathbb{R})$ be diagonal. Then A is symmetric.

(7) A matrix A is normal if and only if A^t is normal.

(8) A normal invertible matrix is orthogonal.

Answers.

(1) **False!** However, the only symmetric and skew-symmetric matrix at the same time is the zero matrix.

(2) **False and never!** Recall that a matrix $A = (a_{ij}) \in M_n(\mathbb{R})$ is skew-symmetric if it satisfies $A = -A^t$. For an $n \times n$ matrix, this condition implies that the diagonal elements must all be zero since $a_{ii} = -a_{ii}$ leads to $2a_{ii} = 0$, hence $a_{ii} = 0$. Additionally, the off-diagonal elements satisfy $a_{ij} = -a_{ji}$. For example, the general form of a 3×3 skew-symmetric matrix is:

$$A = \begin{pmatrix} 0 & a & b \\ -a & 0 & c \\ -b & -c & 0 \end{pmatrix},$$

where $a, b, c \in \mathbb{R}$.

(3) **True.** Here is a proof:

$$0 = AB \implies 0 = 0^t = (AB)^t = B^t A^t = BA.$$

(4) **True.** Let A be an invertible matrix, i.e., $A^{-1}A = I$. Then

$$I = I^t = (A^{-1}A)^t = A^t(A^{-1})^t = A(A^{-1})^t.$$

So, $(A^{-1})^t = A^{-1}$, which signifies that A^{-1} is symmetric.

(5) **True.** A proof may be consulted in Solution 5.3.1.

(6) **True.** Indeed, we have:

$$\begin{pmatrix} a_1 & 0 & \cdots & 0 \\ 0 & a_2 & \ddots & \vdots \\ \vdots & \ddots & \ddots & \\ 0 & \cdots & 0 & a_n \end{pmatrix}^t = \begin{pmatrix} a_1 & 0 & \cdots & 0 \\ 0 & a_2 & \ddots & \vdots \\ \vdots & \ddots & \ddots & \\ 0 & \cdots & 0 & a_n \end{pmatrix}.$$

(7) **True.** The easy proof is left to the reader.

(8) **False!** Consider $A = \begin{pmatrix} 1 & 0 \\ 0 & 2 \end{pmatrix}$. Then, A is symmetric (hence normal!) and invertible. However,

$$A^t A = \begin{pmatrix} 1 & 0 \\ 0 & 2 \end{pmatrix} \begin{pmatrix} 1 & 0 \\ 0 & 2 \end{pmatrix} = \begin{pmatrix} 1 & 0 \\ 0 & 4 \end{pmatrix} \neq \begin{pmatrix} 1 & 0 \\ 0 & 1 \end{pmatrix},$$

that is, A is not orthogonal.

5.3. Exercises with Solutions

Exercise 5.3.1. Let $A \in M_n(\mathbb{R})$. Which of the expressions $A + A^t$, $A^t A$, $A A^t$, $A^2 A^t$, and $A - A^t$ are symmetric? Skew-symmetric?

Solution 5.3.1. Since

$$(A + A^t)^t = A^t + (A^t)^t = A^t + A = A + A^t,$$

$A + A^t$ is symmetric. Next, both of $A^t A$ and $A A^t$ are symmetric (where, here, A could be a rectangular matrix); for example, $A A^t$ is symmetric because

$$(A A^t)^t = (A^t)^t A^t = A A^t.$$

Regarding $A^2 A^t$, it is not symmetric, and we need to give a counterexample. Take $A = \begin{pmatrix} 0 & 1 \\ 2 & 0 \end{pmatrix}$, and so

$$A^2 A^t = \begin{pmatrix} 0 & 1 \\ 2 & 0 \end{pmatrix} \begin{pmatrix} 0 & 1 \\ 2 & 0 \end{pmatrix} \begin{pmatrix} 0 & 2 \\ 1 & 0 \end{pmatrix} = \begin{pmatrix} 0 & 4 \\ 2 & 0 \end{pmatrix},$$

which reveals that $A^2 A^t$ is not symmetric. In the end, $A - A^t$ is not symmetric, as if, e.g., $A = \begin{pmatrix} 0 & 1 \\ 0 & 0 \end{pmatrix}$, then $A - A^t = \begin{pmatrix} 0 & 1 \\ -1 & 0 \end{pmatrix}$, which is not symmetric.

Out of the given expressions, only $A - A^t$ is skew-symmetric. Indeed,

$$(A - A^t)^t = A^t - (A^t)^t = A^t - A = -(A - A^t).$$

However, the only matrix that is both symmetric and skew-symmetric at the same time is the zero matrix. If $A + A^t$, $A^t A$, or $A A^t$ were skew-symmetric, they would have to be zero. However, this is not always the case. For $A + A^t$, any matrix where A^t is not equal to $-A$ (for example, $A = I$) serves as a counterexample. For $A^t A$ and $A A^t$, their nullity implies that A must be zero (see Exercise 1.3.46), so any non-zero matrix is a counterexample. Finally, for $A^2 A^t$, the same counterexample used above, i.e., $A = \begin{pmatrix} 0 & 1 \\ 2 & 0 \end{pmatrix}$, can be used to show that $A^2 A^t$ is not skew-symmetric.

Exercise 5.3.2. Let $A, B \in M_n(\mathbb{R})$.

(1) Show that if A is symmetric, so are A^n, $A + I$. What if "symmetric" is replaced with "skew-symmetric"?
(2) Suppose here that A and B are symmetric.
   (a) Show that $\alpha A + \beta B$ is symmetric for all real α and β.
   (b) Show that $AB + BA$ is symmetric and that $AB - BA$ is skew-symmetric.
   (c) Is AB always symmetric?
   (d) Show that AB is symmetric if and only if $AB = BA$.
   (e) Show that ABA is symmetric.
(3) Suppose now that A and B are skew-symmetric.
   (a) Show that $\alpha A + \beta B$ is skew-symmetric for all real α and β.
   (b) Is AB skew-symmetric?
   (c) Show that AB is skew-symmetric if and only if $AB = -BA$.
   (d) Show that ABA is skew-symmetric.

Solution 5.3.2.

(1) If A is symmetric, meaning that $A^t = A$, we have

$$(A^n)^t = (\underbrace{AA \cdots A}_{n \text{ times}})^t = (\underbrace{A^t A^t \cdots A^t}_{n \text{ times}}) = (\underbrace{AA \cdots A}_{n \text{ times}}) = A^n,$$

which indicates that A^n is symmetric. The matrix $A + I$ is symmetric because

$$(A + I)^t = A^t + I^t = A + I.$$

Now, assuming $A^t = -A$. In this scenario, A^n is not always skew-symmetric; the outcome depends on the parity of n. Specifically, when n is odd, we can express

$$(A^n)^t = (\underbrace{A^t A^t \cdots A^t}_{n \text{ times}}) = (-1)^n (\underbrace{AA \cdots A}_{n \text{ times}}) = -A^n,$$

showing that A^n is skew-symmetric provided n is odd. If n is even, for example $n = 2$, then A^2 is not necessarily skew-symmetric when A is. For example, $A = \begin{pmatrix} 0 & 1 \\ -1 & 0 \end{pmatrix}$ is skew-symmetric, but

$$A^2 = \begin{pmatrix} 0 & 1 \\ -1 & 0 \end{pmatrix} \begin{pmatrix} 0 & 1 \\ -1 & 0 \end{pmatrix} = \begin{pmatrix} -1 & 0 \\ 0 & -1 \end{pmatrix}$$

is not skew-symmetric.

In fact, if n is even, so $(-1)^n = 1$, then A^n is always symmetric if A is skew-symmetric for

$$(A^n)^t = (\underbrace{A^t A^t \cdots A^t}_{n \text{ times}}) = (-1)^n (\underbrace{AA \cdots A}_{n \text{ times}}) = A^n.$$

Ultimately, $A + I$ is not skew-symmetric if A is skew-symmetric. Indeed, even in the simplest case of $A = 0$, which is skew-symmetric, $A + I = I$ is not skew-symmetric.

(2) (a) Let $\alpha, \beta \in \mathbb{R}$. We have

$$(\alpha A + \beta B)^t = (\alpha A)^t + (\beta B)^t = \alpha A^t + \beta B^t = \alpha A + \beta B,$$

which shows that $\alpha A + \beta B$ is symmetric.

(b) Since

$$(AB + BA)^t = (AB)^t + (BA)^t = B^t A^t + A^t B^t$$
$$= BA + AB = AB + BA,$$

we see that $AB + BA$ is symmetric. Similarly,

$$(AB - BA)^t = (AB)^t - (BA)^t$$
$$= B^t A^t - A^t B^t = BA - AB = -(AB - BA)$$

demonstrates that $AB - BA$ is skew-symmetric in the case both A and B are symmetric.

(c) The product AB is not always symmetric when both A and B are symmetric. As a counterexample, consider the symmetric matrices

$$A = \begin{pmatrix} -1 & 0 \\ 0 & 2 \end{pmatrix} \text{ and } B = \begin{pmatrix} 0 & 1 \\ 1 & 0 \end{pmatrix}.$$

Then, $AB = \begin{pmatrix} 0 & -1 \\ 2 & 0 \end{pmatrix}$ is not symmetric.

(d) Since $(AB)^t = B^t A^t = BA$, it becomes clear that AB is symmetric if and only if $AB = BA$.

(e) The observation

$$(ABA)^t = [(AB)A]^t = A^t(AB)^t = A^t B^t A^t = ABA$$

confirms that ABA is a symmetric matrix whenever A and B are.

(3) (a) Let $\alpha, \beta \in \mathbb{R}$. We have

$$(\alpha A + \beta B)^t = (\alpha A)^t + (\beta B)^t$$
$$= \alpha A^t + \beta B^t$$
$$= -\alpha A - \beta B$$
$$= -(\alpha A + \beta B),$$

which means that $\alpha A + \beta B$ is skew-symmetric.

(b) In general, AB need not be skew-symmetric if A and B are skew-symmetric. For instance, take the skew-symmetric matrices $A = \begin{pmatrix} 0 & -1 \\ 1 & 0 \end{pmatrix}$ and $B = \begin{pmatrix} 0 & 1 \\ -1 & 0 \end{pmatrix}$. Then $AB = \begin{pmatrix} 1 & 0 \\ 0 & 1 \end{pmatrix}$ is not skew-symmetric.

(c) Since $(AB)^t = B^t A^t = (-B)(-A) = BA$, it becomes apparent that AB is skew-symmetric if and only if $AB = -BA$.

(d) Given
$$(ABA)^t = A^t(AB)^t = A^t B^t A^t = (-A)(-B)(-A),$$
we can see that $(ABA)^t = -ABA$, i.e., ABA is skew-symmetric, as required.

Exercise 5.3.3. Find a non-symmetric matrix $A \in M_n(\mathbb{R})$ such that A^2 is symmetric.

Solution 5.3.3. Let A be defined by:
$$A = \begin{pmatrix} a & 1 \\ 0 & -a \end{pmatrix},$$
where $a \in \mathbb{R}$. Clearly, A is not symmetric but
$$A^2 = \begin{pmatrix} a^2 & 0 \\ 0 & a^2 \end{pmatrix},$$
is symmetric.

Exercise 5.3.4. (Cf. Exercise 5.3.12) Let $A \in M_n(\mathbb{R})$ be a symmetric matrix.

(1) Suppose that $A^2 = 0$. Show that $A = 0$.
(2) Deduce that if $A^p = 0$ for some $p \geq 2$, then $A = 0$.

Remark. See Exercise 5.3.19 for another proof.

Solution 5.3.4.

(1) Since $A^2 = 0$, it is clear that $\operatorname{tr}(A^2) = 0$. Also, since A is symmetric, that is, $A^t = A$, we too have $\operatorname{tr}(A^t A) = 0$. By Exercise 1.3.46, we conclude that $A = 0$, as needed.

(2) We claim that $A^p = 0$ always yields $A^{p-1} = 0$ as long as $p - 1 \geq 1$. Indeed, we have
$$(A^{p-1})^2 = A^{2p-2} = A^{p+p-2} = A^p A^{p-2} = 0.$$

Since A^{p-1} is symmetric, the previous question implies $A^{p-1} = 0$. By continuing this, we can eventually show that $A = 0$, as desired.

Exercise 5.3.5. Let $T \in M_n$ be such that $T^p = 0$ for some $p \in \mathbb{N}$, with $p \geq 2$. Show that T^{p-1} is symmetric if and only if $T^{p-1} = 0$.

Solution 5.3.5. We show only the nontrivial implication: If T^{p-1} is symmetric, then $T^{p-1} = 0$. Since $T^p = 0$, it follows that

$$(T^{p-1})^2 = T^{2p-2} = T^p T^{p-2} = 0,$$

which, with Exercise 5.3.4, yields $T^{p-1} = 0$, as desired.

Exercise 5.3.6. Give the general form of real normal 2×2 matrices.

Solution 5.3.6. Let

$$A = \begin{pmatrix} a & b \\ c & d \end{pmatrix},$$

where $a, b, c, d \in \mathbb{R}$. Then

$$A^t = \begin{pmatrix} a & c \\ b & d \end{pmatrix},$$

so that

$$AA^t = \begin{pmatrix} a^2 + b^2 & ac + bd \\ ac + bd & c^2 + d^2 \end{pmatrix} \text{ and } A^t A = \begin{pmatrix} a^2 + c^2 & ab + cd \\ ab + cd & b^2 + d^2 \end{pmatrix}.$$

In order that A be normal, the entries must obey

$$b^2 - c^2 = 0 \text{ and } (b - c)(a - d) = 0.$$

So basically there are two cases to look at:

- If $b = c$, then A is clearly symmetric (hence normal) regardless of the other values.
- If $b = -c$, then $a = d$.

Accordingly, the 2×2 real normal matrices are either *symmetric* or of the form

$$A = \begin{pmatrix} a & b \\ -b & a \end{pmatrix}.$$

Exercise 5.3.7. Let $A \in M_n(\mathbb{R})$ be normal. Show that A^p is normal for all $p \in \mathbb{N}$.

Solution 5.3.7. Let $p \in \mathbb{N}$ and let A be normal, i.e., $A^t A = AA^t$. Applying the result of Exercise 1.3.8 (with A^t playing the role of B there), gives

$$(A^p)^t A^p = (A^t)^p A^p = (A^t A)^p = (AA^t)^p = A^p (A^t)^p = A^p (A^p)^t,$$

indicating that A^p is normal.

Exercise 5.3.8. Let $A \in M_n(\mathbb{R})$ be a matrix such that $A + I$ is orthogonal. Show that A is normal.

Solution 5.3.8. Since $A+I$ is orthogonal, we have $(A+I)^t(A+I) = I$, or $A^tA + A^t + A + I = I$. Therefore, $A^tA = -A^t - A$, or equivalently, $(-A)^t(-A) = (-A)^t + (-A)$. By Exercise 1.3.25, $(-A)$ commutes with $(-A)^t$, or simply $A^tA = AA^t$, which shows the normality of A.

Exercise 5.3.9. Let $T \in M_n(\mathbb{R})$.

(1) Show that there are $A, B \in M_n(\mathbb{R})$, where A is symmetric and B is skew-symmetric, such that $T = A + B$. Necessarily, $A = (T + T^t)/2$ and $B = (T - T^t)/2$.

   Remark. The decomposition $T = A + B$, where A is symmetric and B is skew-symmetric, is better known as the Cartesian (or Toeplitz) decomposition of matrix T.

(2) Let $T = A + B$, where A is symmetric and B is skew-symmetric. Show that T is normal if and only if $AB = BA$.

Solution 5.3.9.

(1) Suppose A is symmetric and B is skew-symmetric, meaning that $A^t = A$ and $B^t = -B$. Since $T = A + B$, it follows that $T^t = A^t + B^t = A - B$. Thus, $T + T^t = 2A$ and $T - T^t = 2B$. Therefore, $A = (T + T^t)/2$ and $B = (T - T^t)/2$.

   Conversely, if $A = (T + T^t)/2$ and $B = (T - T^t)/2$, then A is symmetric and B is skew-symmetric, and

   $$T = \frac{T + T^t}{2} + \frac{T - T^t}{2}.$$

(2) Since $T = A + B$ and $T^t = A - B$, we can write

   $$T^tT = (A - B)(A + B) = A^2 + AB - BA - B^2$$

   and

   $$TT^t = (A + B)(A - B) = A^2 - AB + BA - B^2.$$

   Therefore, T is normal if and only if $AB - BA = -AB + BA$, that is, if and only if $2AB = 2BA$, or equivalently, $AB = BA$.

Exercise 5.3.10. Let $T \in M_n(\mathbb{R})$ be expressed as $T = A + B$, where A is symmetric and B is skew-symmetric.

(1) Show that $T = 0$ if and only if $A = B = 0$.

(2) By an example, show that the condition A being symmetric and B being skew-symmetric cannot just be dropped.

Solution 5.3.10.

(1) If $A = B = 0$, then it is plain that $T = 0$. Now, let us assume that $0 = T = A + B$. Thus, $0 = T^t = A - B$, and so $A = (T + T^t)/2 = 0$. Also, $B = (T - T^t)/2 = 0$, as needed.

(2) Let $A = \begin{pmatrix} 0 & 1 \\ 0 & 0 \end{pmatrix}$ and $B = \begin{pmatrix} 0 & -1 \\ 0 & 0 \end{pmatrix}$. Then $T := A + B = \begin{pmatrix} 0 & 0 \\ 0 & 0 \end{pmatrix}$. Note that neither A is symmetric nor B is skew-symmetric, and we can observe that $A \neq 0$ and $B \neq 0$.

Exercise 5.3.11. Let $A, B, C, D \in M_n(\mathbb{R})$ be such that $A + B = C + D$, where A, C are symmetric and B, D are skew-symmetric. Show that $A = C$ and $B = D$.

Solution 5.3.11. The hypothesis $A + B = C + D$ may be rewritten as $A - C = D - B$. Furthermore, since $A - C$ is symmetric because A and C are, and since $D - B$ is skew-symmetric because B and D are, the previous exercise suggests that $A = C$ and $B = D$, as wished.

Exercise 5.3.12. Suppose $A \in M_n(\mathbb{R})$ is a normal matrix such that $A^2 = 0$. Show that $A = 0$. Deduce that if $A^p = 0$ for some $p \geq 2$, where A is normal, then $A = 0$.

Solution 5.3.12. Given $A^2 = 0$, we have $A^t A^2 A^t = A^t 0 A^t = 0$, and so $A^t A A A^t = 0$. Since A is normal, it follows that $A^t A A^t A = 0$. In other words, $(A^t A)^2 = 0$. Since $A^t A$ is symmetric, Exercise 5.3.4 yields $A^t A = 0$, thereupon $A = 0$ by referring to, e.g., Exercise 1.3.46.

The general case is very similar to that of Exercise 5.3.4, we leave details to readers.

Exercise 5.3.13. Let $A, B \in M_n(\mathbb{R})$ be normal matrices that obey $AB = 0$. Show that $BA = 0$.

Solution 5.3.13. Suppose $AB = 0$. Since $A^t A = AA^t$ and $B^t B = BB^t$, we can write

$$B^t B A A^t = BB^t A^t B = B(AB)^t A = B0A = 0.$$

Thus,

$$(BA)^t BA (BA)^t = A^t B^t B A A^t B^t = A^t 0 B^t = 0.$$

By Exercise 1.3.46, we conclude that $BA = 0$.

Exercise 5.3.14. Let A and B be two normal matrices of the same size.

(1) Is AB normal?
(2) Is $A + B$ normal?
(3) Assume here that $AB^t = B^t A$. Show that AB and $A + B$ are normal.

> **Remark**. In light of Exercise 5.3.26 below, the hypothesis $AB^t = B^t A$ may be replaced with $AB = BA$.

Solution 5.3.14.

(1) The answer is negative. Consider $A = \begin{pmatrix} -1 & 0 \\ 0 & 2 \end{pmatrix}$ and $B = \begin{pmatrix} 0 & 1 \\ 1 & 0 \end{pmatrix}$, which are symmetric, in particular, normal. However, $AB = \begin{pmatrix} 0 & -1 \\ 2 & 0 \end{pmatrix}$ is not normal, either by directly checking that $(AB)^t AB \neq AB(AB)^t$, or by observing that the matrix AB is not in the form of normal 2×2 matrices (see Solution 5.3.6).

(2) The answer is no. Consider the matrices A and B defined as

$$A = \begin{pmatrix} 1 & -1 \\ 1 & 1 \end{pmatrix}, \ B = \begin{pmatrix} -1 & 2 \\ 2 & 1 \end{pmatrix},$$

which are normal, as readers may check (B is even symmetric). However, $A + B = \begin{pmatrix} 0 & 1 \\ 3 & 2 \end{pmatrix}$ is not normal.

(3) Since $AB^t = B^t A$, we have $(AB^t)^t = (B^t A)^t$, that is, $BA^t = A^t B$. Besides, $A^t A = AA^t$ and $B^t B = BB^t$. Next, we can write

$$\begin{aligned}
(AB)^t AB &= B^t A^t AB \\
&= B^t A A^t B \\
&= AB^t A^t B \\
&= AB^t B A^t \\
&= ABB^t A^t \\
&= AB(AB)^t,
\end{aligned}$$

which demonstrates that AB is normal.

In order to show that $A + B$ is normal, we need to verify that $(A + B)(A + B)^t = (A + B)^t(A + B)$.

Expanding $(A+B)(A+B)^t$, we get $AA^t+AB^t+BA^t+BB^t$, and expanding $(A+B)^t(A+B)$, we get $A^tA+A^tB+B^tA+B^tB$. Since A and B are normal, $AA^t = A^tA$ and $BB^t = B^tB$. Therefore, we only need to compare $AB^t + BA^t$ and $A^tB + B^tA$. Since $AB^t = B^tA$ and $BA^t = A^tB$, we can see that $AB^t + BA^t = A^tB + B^tA$.

This says that $(A+B)(A+B)^t = (A+B)^t(A+B)$, proving that $A + B$ is normal.

Exercise 5.3.15. Show that symmetric and skew-symmetric matrices are normal, but a normal matrix need not be symmetric (or skew-symmetric).

Solution 5.3.15. Let $A \in M_n(\mathbb{R})$. If A is symmetric, i.e., $A^t = A$, then

$$A^tA = A^2 = AA^t,$$

which shows that A is normal. If, now, A is skew-symmetric, meaning that $A^t = -A$, then

$$A^tA = (-A)A = -A^2 = A(-A) = AA^t,$$

which also says that A is normal.

If A is normal, then A need not be symmetric (or skew-symmetric). For example, readers may check that $A = \begin{pmatrix} 0 & 1 \\ -1 & 0 \end{pmatrix}$ is normal, but it is not symmetric. Also, $A = \begin{pmatrix} 1 & 0 \\ 0 & 1 \end{pmatrix}$ is normal without being skew-symmetric.

Let us supply an example, borrowed from [21], of a normal matrix that is neither symmetric nor skew-symmetric (it is not orthogonal either). Let

$$A = \begin{pmatrix} 1 & 1 & 0 \\ 0 & 1 & 1 \\ 1 & 0 & 1 \end{pmatrix}.$$

Clearly, A is not symmetric, and it is not skew-symmetric as it has at least one nonzero element on its main diagonal. However,

$$A^tA = \begin{pmatrix} 2 & 1 & 1 \\ 1 & 2 & 1 \\ 1 & 1 & 2 \end{pmatrix} = AA^t,$$

signifying that A is normal.

Exercise 5.3.16. Let $A, B \in M_n(\mathbb{R})$ be two orthogonal matrices.
(1) Is AB orthogonal?
(2) Show that A^p is orthogonal for all $p \in \mathbb{N}$.
(3) Is $A + B$ orthogonal?

Solution 5.3.16.

(1) The answer is yes. By assumption, $A^t A = I$ and $B^t B = I$, whereby

$$(AB)^t AB = B^t A^t AB = B^t IB = B^t B = I,$$

indicating that AB is orthogonal.

(2) Let $p \in \mathbb{N}$. We proceed by induction on p. The base case is true by assumption. Now, assume A^p is orthogonal, and so by the previous question, $A^p A$, or equivalently A^{p+1}, is orthogonal.

(3) The answer is negative, even when $A = B$! If $A = B$ is orthogonal, then $A + B = 2A$ is not orthogonal because

$$(A + B)^t (A + B) = 2A^t(2A) = 4A^t A = 4I \neq I.$$

Exercise 5.3.17. Let $A, B \in M_n(\mathbb{R})$. Assume that $P^t AP = B$, where P is orthogonal.
(1) Show that A is symmetric if and only if B is.
(2) Show that A is normal if and only if B is.

Solution 5.3.17.

(1) Assume that A is symmetric. Then

$$B^t = (P^t AP)^t = P^t(P^t A)^t = P^t A^t P^{tt} = P^t AP = B,$$

showing that B is symmetric. Conversely, if B is symmetric and since $PP^t = I$, we may write $A = PBP^t$, so that we may check, as before, that A is symmetric.

(2) Given that $P^t AP = B$ with P orthogonal ($P^t P = I$), we show that A is normal if and only if B is.

Since A is normal, we have $AA^t = A^t A$. Multiplying both sides on the left by P^t and on the right by P, we obtain $P^t AA^t P = P^t A^t AP$, or equivalently, $P^t APP^t A^t P = P^t A^t PP^t AP$. Using $P^t AP = B$, this simplifies to $BB^t = B^t B$, showing that B is normal.

Conversely, if B is normal, i.e., $BB^t = B^t B$, substituting $B = P^t AP$ and reversing the argument gives $AA^t = A^t A$. Thus, A is normal if and only if B is normal.

Exercise 5.3.18. Let $A, B \in M_n$, where A is normal.
(1) Show that $\ker(A) = \ker(A^t)$.
(2) Give an example of a non-normal B such that $\ker(B) = \ker(B^t)$.

Solution 5.3.18.

(1) By Exercise 3.3.14, $\ker(A) = \ker(A^t A)$. So,

$$\ker(A^t) = \ker[(A^t)^t A^t] = \ker(AA^t).$$

But, A is normal, i.e., $AA^t = A^t A$; thus,

$$\ker(A) = \ker(A^t A) = \ker(AA^t) = \ker(A^t).$$

(2) Take $B = \begin{pmatrix} 0 & 1 \\ 2 & 0 \end{pmatrix}$. Then, B is invertible, and so $\ker B = \ker(B^t) = \{(0,0)\}$, while B is not normal.

Exercise 5.3.19. Let $A \in M_n(\mathbb{R})$ be normal and $p \in \mathbb{N}$.
(1) Show that

$$\ker(A) = \ker(A^2) = \cdots = \ker(A^p).$$

(2) Infer that if A is nilpotent, then $A = 0$.

Solution 5.3.19.

(1) It is clear that $\ker A \subset \ker(A^2)$, as $Ax = 0$ implies $A^2 x = 0$. Now, let $x \in \ker A^2$, which means $A^2 x = 0$. So, $Ax \in \ker A = \ker(A^t)$ because A is normal. Therefore, $A^t Ax = 0$. Using the normality of A again, we can see that $x \in \ker(A^t A) = \ker A$. In other words, we have shown that $\ker(A^2) \subset \ker A$, which leads to $\ker(A) = \ker(A^2)$.

Now, if $x \in \ker A$, it means that $Ax = 0$. This condition further implies that $A^p x = 0$ for all $p \geq 2$. As a result, we have the following nested relationship:

$$\ker(A) \subset \ker(A^2) \subset \cdots \subset \ker(A^p).$$

To demonstrate that $\ker(A^3) \subset \ker(A^2)$, let us assume $x \in \ker(A^3)$, meaning $A^3 x = 0$, which can be rewritten as $A^2 Ax = 0$, resulting in $Ax \in \ker(A^2) = \ker(A)$. Therefore, $A^2 x = 0$, or equivalently, $x \in \ker(A^2)$. In other words, we have shown that $\ker(A^3) \subset \ker(A^2) \subset \ker A$. By "downward" induction, it may be shown that $\ker(A) \supset \ker(A^2) \supset \cdots \supset \ker(A^p)$. Consequently, we have established the required equalities

$$\ker(A) = \ker(A^2) = \cdots = \ker(A^p).$$

(2) Supposing $A^p = 0$ for some p means $A^p x = 0$ for all x, or $\ker(A^p) = \mathbb{R}^n$ (assuming A is defined on $\mathbb{R}^n$). By the previous question, $\ker A = \mathbb{R}^n$, so $Ax = 0$ for all x, implying $A = 0$.

Exercise 5.3.20. Let A be a square matrix that satisfies $A = A^t A$. Show that $A = A^t$, hence $A^2 = A$. Is the converse true?

Solution 5.3.20. We have
$$A^t = (A^t A)^t = A^t (A^t)^t = A^t A = A.$$
So, $A = A^t A = (A^t)^t A = AA = A^2$, as wished.

The converse is untrue. To see it, consider $A = \begin{pmatrix} 1 & 2025 \\ 0 & 0 \end{pmatrix}$. Then,
$$A^2 = \begin{pmatrix} 1 & 2025 \\ 0 & 0 \end{pmatrix} \begin{pmatrix} 1 & 2025 \\ 0 & 0 \end{pmatrix} = \begin{pmatrix} 1 & 2025 \\ 0 & 0 \end{pmatrix} = A,$$
yet $A^t = \begin{pmatrix} 1 & 0 \\ 2025 & 0 \end{pmatrix} \neq A$.

Exercise 5.3.21. Let $A \in M_n(\mathbb{R})$ satisfy $\operatorname{tr}(A^t A) = \operatorname{tr}(A^2)$. Show that $A^t = A$.

Solution 5.3.21. First, observe that $\operatorname{tr}(A^2) = \operatorname{tr}[(A^2)^t] = \operatorname{tr}[(A^t)^2]$. Let us now compute $\operatorname{tr}[(A - A^t)^2]$ for a reason that will soon be clear to readers. We have
$$\operatorname{tr}(A - A^t)^2 = \operatorname{tr}[A^2 - AA^t - A^t A + (A^t)^2]$$
$$= \operatorname{tr}(A^2) - \operatorname{tr}(AA^t) - \operatorname{tr}(A^t A) + \operatorname{tr}((A^t)^2).$$

Since $\operatorname{tr}(AA^t) = \operatorname{tr}(A^t A)$ and $\operatorname{tr}(A^t A) = \operatorname{tr}(A^2) = \operatorname{tr}(A^t)^2$, it ensues that $\operatorname{tr}(A - A^t)^2 = 0$. Since $(A - A^t)^t = A^t - A$, it follows that $\operatorname{tr}[-(A - A^t)^t(A - A^t)] = 0$, or equivalently, $\operatorname{tr}(A - A^t)^t(A - A^t) = 0$. Exercise 1.3.46 thus gives $A^t = A$, as wished.

Remark. In a very similar way, the equation $AA^t = A^2$, where $A \in M_n(\mathbb{R})$, leads to $A^t = A$, and details are left to interested readers.

Exercise 5.3.22. Let $A \in M_n(\mathbb{R})$ satisfy $\operatorname{tr}(A^t A) = -\operatorname{tr}(A^2)$. Show that $A^t = -A$.

Solution 5.3.22. We apply the solution method to Exercise 5.3.21, providing only a brief outline and omitting many details for conciseness. Since $\operatorname{tr}(A^t A) = -\operatorname{tr}[(A^2)^t] = -\operatorname{tr}(A^2)$, it follows that
$$\operatorname{tr}(A + A^t)^2 = \operatorname{tr}(A^2) + \operatorname{tr}(AA^t) + \operatorname{tr}(A^t A) + \operatorname{tr}[(A^t)^2] = 0.$$

Since $(A + A^t)^t = A + A^t$, by invoking Exercise 1.3.46, we see that $A^t = -A$, which completes the proof.

Remark. In a similar manner, the equation $AA^t = -A^2$, where $A \in M_n(\mathbb{R})$, implies $A^t = -A$; further details are left to interested readers.

Exercise 5.3.23. (Cf. Exercise 5.4.6) Let $A \in M_n$ satisfy $\mathrm{tr}[(A^t)^2 A^2] = \mathrm{tr}[(A^t A)^2]$. By computing $\mathrm{tr}[(A^t A - AA^t)(A^t A - AA^t)^t]$ or else, show that A is normal.

Solution 5.3.23. It is easy to check that

$$\mathrm{tr}[(A^t A - AA^t)(A^t A - AA^t)^t] = 2[\mathrm{tr}[(A^t)^2 A^2] - \mathrm{tr}(A^t A)^2] = 0,$$

from which we deduce that $A^t A - AA^t = 0$, or equivalently that A is normal.

Exercise 5.3.24. Let $A \in M_n$ satisfy $A(A^t A) = (A^t A)A$. Show that A is normal.

Solution 5.3.24. Since $A(A^t A) = (A^t A)A$, it follows that

$$\mathrm{tr}[(A^t A)^2] = \mathrm{tr}(A^t AA^t A) = \mathrm{tr}(A^t A^t AA) = \mathrm{tr}[(A^t)^2 A^2],$$

which results in the normality of A thanks to Exercise 5.3.23.

Exercise 5.3.25. Let $T \in M_n(\mathbb{R})$ be such that $T^2 = (T^t + T)/2$.
  (1) Show that T is normal.
  (2) By writing $T = A + B$, where A is symmetric and B is skew-symmetric, transform the equation $T^2 = (T^t + T)/2$ into $R + S = 0$, where R is symmetric and S is skew-symmetric.
  (3) Infer that $B = 0$, and so T is symmetric with $T^2 = T$.

Solution 5.3.25.
  (1) The assumption $T^2 = (T^t + T)/2$ may be rewritten as $2T^2 = T^t + T$, which then simplifies to $T^t = 2T^2 - T$. Hence

$$TT^t = T(2T^2 - T) = 2T^3 - T^2 = (2T^2 - T)T = T^t T,$$

  demonstrating that T is normal.
  (2) Let us express T as $T = A + B$, where A is a symmetric matrix and B is a skew-symmetric matrix. Since T is normal, we know by Exercise 5.3.10 that $AB = BA$. This allows us to rewrite $T^2 = (A + B)^2 = A^2 + 2AB + B^2$. Substituting $T^2 = (T^t + T)/2$, we get $A^2 + 2AB + B^2 = A$, which simplifies to $A^2 + B^2 - A + 2AB = 0$. The term $A^2 + B^2 - A$ is easily seen to be symmetric, whereas $2AB$, or equivalently AB, is skew-symmetric for

$$(AB)^t = B^t A^t = (-B)A = -BA = -AB.$$

(3) Given that $A^2 + B^2 - A + 2AB = 0$, from Exercise 5.3.10, we can deduce that $A^2 + B^2 - A = 0$ and $2AB = 0$. Since $AB = 0$, this implies that $B^3 = AB - A^2B = 0$. Since B is skew-symmetric, it is normal, and so $B = 0$ by Exercise 5.3.12. Consequently, $T = A$ is symmetric, as required.

Exercise 5.3.26. Let $A, B \in M_n$. If either A or B is normal, then
$$AB = BA \iff AB^t = B^t A.$$

Solution 5.3.26. Using properties of the trace allows us to write

$$\operatorname{tr}[(AB - BA)^t(AB - BA)]$$
$$= \operatorname{tr}(B^t A^t AB - B^t A^t BA - A^t B^t AB + A^t B^t BA)$$
$$= \operatorname{tr}(B^t A^t AB) - \operatorname{tr}(B^t A^t BA) - \operatorname{tr}(A^t B^t AB) + \operatorname{tr}(A^t B^t BA)$$
$$= \operatorname{tr}(A^t ABB^t) - \operatorname{tr}(AB^t A^t B) - \operatorname{tr}(BA^t B^t A) + \operatorname{tr}(AA^t B^t B)$$
$$= \operatorname{tr}(A^t ABB^t) - \operatorname{tr}(AB^t A^t B) - \operatorname{tr}(ABA^t B^t) + \operatorname{tr}(AA^t B^t B)$$
$$= \operatorname{tr}(A^t ABB^t - AB^t A^t B - ABA^t B^t + AA^t B^t B).$$

Replacing B with its transpose leads to

$$\operatorname{tr}[(AB^t - B^t A)^t(AB^t - B^t A)]$$
$$= \operatorname{tr}(A^t AB^t B - ABA^t B^t - AB^t A^t B + AA^t BB^t).$$

Thus,

$$\operatorname{tr}[(AB - BA)^t(AB - BA)] - \operatorname{tr}[(AB^t - B^t A)^t(AB^t - B^t A)]$$
$$= \operatorname{tr}(-A^t AB^t B + A^t ABB^t + AA^t B^t B - AA^t BB^t)$$
$$= \operatorname{tr}[-(A^t A - AA^t)(B^t B - BB^t)]$$
$$= -\operatorname{tr}[(A^t A - AA^t)(B^t B - BB^t)].$$

If either B or A is normal, then the last term in the previous equalities is null, so

$$\operatorname{tr}[(AB - BA)^t(AB - BA)] = \operatorname{tr}[(AB^t - B^t A)^t(AB^t - B^t A)]$$

from which we get

$$AB = BA \iff AB^t = B^t A,$$

as suggested.

Remark. The result of this exercise serves as the foundation for a much deeper theorem. If you have the necessary background, you may refer to [17] for further exploration.

5.4. Exercises without Solutions

Exercise 5.4.1. Find the values of the scalars a, b, and c that make the matrix

$$\begin{pmatrix} 2 & a - 2b + 2c & 2a + b + c \\ 3 & 5 & a + c \\ 0 & -2 & 7 \end{pmatrix}$$

symmetric.

Exercise 5.4.2. Show that the identity matrix I_2 possesses infinitely many *symmetric* square roots.

Exercise 5.4.3. Let $q \in \mathbb{R}$. Can we find a non-normal matrix A such that

$$AA^t = qA^tA?$$

Exercise 5.4.4. (Cf. [16]) Is it true that if $A, B, C \in M_n$ are symmetric and ABC is symmetric, then at least two of A, B, and C must commute?

Exercise 5.4.5. Find a matrix $A \in M_n$ such that

$$AA^t + A^tA = I, \ A^2 = (A^t)^2 = 0.$$

Can such a matrix be normal?

Exercise 5.4.6. Let $A, B \in M_n(\mathbb{R})$ be both symmetric. Using Exercise 5.3.21, or else, show that $\operatorname{tr}[(AB)^2] = \operatorname{tr}(A^2B^2)$ if and only if $AB = BA$.

Exercise 5.4.7. Let p be a real polynomial and $A \in M_n(\mathbb{R})$. Show that $[p(A)]^t = p(A^t)$. In particular, $p(A)$ is symmetric if A is symmetric.

Exercise 5.4.8. Let $A, B \in M_n(\mathbb{C})$.

(1) Assume that $\operatorname{tr}(A^*A) = 0$, where A^* is defined in the second remark below Definition 5.1.1. Show that $A = 0$.
(2) Infer that each of the equations $A^*A = 0$ or $AA^* = 0$ implies $A = 0$.
(3) If $AA^*A = 0$, does it follow that $A = 0$?
(4) Suppose $\operatorname{tr}(XA) = \operatorname{tr}(XB)$ for all $X \in M_n(\mathbb{C})$. Infer that $A = B$.

CHAPTER 6

Eigenvalues

6.1. Course Summary

Throughout this chapter, E designates a vector space over a field $\mathbb{F}$ (which is often $\mathbb{R}$ or $\mathbb{C}$) and $f : E \to E$ is an endomorphism.

DEFINITION 6.1.1. We say that $\lambda \in \mathbb{F}$ is an eigenvalue of f if there exists a nonzero vector $x \in E$ such that if $f(x) = \lambda x$. Any such vector x is called an eigenvector corresponding to λ.

If λ is an eigenvalue of f, then $\ker(f - \lambda I)$ is called the eigenspace of f associated with λ.

Remark. On several occasions, we denote the set of eigenvalues by σ_p, adopting this notation from a more advanced course.

Remark. If $I : E \to E$ is the identity mapping, then λ of f is an eigenvalue of f if and only if $\ker(f - \lambda I) \neq \{0\}$, that is, if and only if $f - \lambda I$ is not one-to-one.

Remark. In some older texts, readers may come across the terms "proper values" and "proper vectors." These are alternative names for eigenvalues and eigenvectors, respectively. These terms arise from translations of the German word "eigen," meaning "proper" or "characteristic."

Remark. We typically denote the eigenspace corresponding to an eigenvalue λ by E_λ. For instance, E_1 is the eigenspace associated with $\lambda = 1$, E_{-2} corresponds to $\lambda = -2$, and so on.

Next, we have an equivalent definition of eigenvalues in the case of finite-dimensional spaces.

THEOREM 6.1.1. *Let $A \in M_n$. Then, λ is an eigenvalue of A if and only if $\det(A - \lambda I) = 0$.*

DEFINITION 6.1.2. The expression $p_A(\lambda) := \det(A - \lambda I_n)$ is a polynomial of degree n in λ, known as the characteristic polynomial of A. The characteristic equation of A is the equation $p_A(\lambda) = 0$.

Remark. There is not much difference in considering $\lambda I - f$ instead of $f - \lambda I$. More precisely, whether we use $\lambda I - f$ or $f - \lambda I$, both forms are equivalent for the purpose of finding eigenvalues because they both lead to the same characteristic equation $\det(A - \lambda I) = 0$, or equivalently, $\det(\lambda I - A) = 0$.

The characteristic polynomial takes an interesting form provided below.

THEOREM 6.1.2. *The characteristic polynomial of $A \in M_n$ takes the form:*

$$p_A(\lambda) = (-1)^n \lambda^n + (-1)^{n-1}(\operatorname{tr} A)\lambda^{n-1} + \cdots + \det A,$$

where the remaining coefficients are numbers without specific names.

Remark. In the special case where $n = 2$, the preceding theorem states that the characteristic polynomial of A is given by:

$$p_A(\lambda) = \lambda^2 - (\operatorname{tr} A)\lambda + \det A.$$

This provides the most efficient way to compute the characteristic polynomial of 2×2 matrices. A proof may be consulted in Exercise 6.3.7.

DEFINITION 6.1.3. Let $A \in M_n$ and λ be an eigenvalue of A.

(1) The algebraic multiplicity of λ is the number of times λ appears as a root of the characteristic polynomial of A.
(2) The geometric multiplicity of λ is the dimension of the eigenspace corresponding to λ, i.e., $\dim \ker(A - \lambda I_n)$.

EXAMPLES 6.1.1.

(1) Let E be a vector space, possibly infinite-dimensional and consider $I : E \to E$ defined by $I(x) = x$. Then the unique eigenvalue of I is $\lambda = 1$.
(2) The zero matrix (or endomorphism) has only one eigenvalue, namely $\lambda = 0$.
(3) The matrix $A = \begin{pmatrix} 0 & 1 \\ -1 & 0 \end{pmatrix}$ has no eigenvalues when defined on $\mathbb{R}^2$, but it has two eigenvalues, $\lambda = \pm i$, when defined over $\mathbb{C}^2$.
(4) Similarly, $A = \begin{pmatrix} 0 & 1 \\ 2 & 0 \end{pmatrix}$ has no eigenvalues when considered over $\mathbb{Q}^2$, as the only possible eigenvalues $\pm\sqrt{2}$, are not rational.

6.2. True or False

Questions. Answer the following questions, providing justifications for your answers:

(1) Let I be the identity matrix in M_n and $p(x) = x - 1$. Then $p(I) = 0_{M_n}$. According to Theorem 6.1.2, it should follow that $\det I = -1$, which is absurd (for $\det I = 1$). What is wrong with this argument?

(2) Let λ be an eigenvalue of f and $\ker(f - \lambda I)$ be its associated eigenspace. Then, all elements of $\ker(f - \lambda I)$ are eigenvectors of f.

(3) A matrix having a null eigenvalue is invertible.

(4) A square matrix of order n, has always n eigenvalues counting the multiplicity of each eigenvalue.

(5) Let A and B be two matrices of the same order. Then AB and BA have the same eigenvalues.

(6) Each real square matrix admits a real eigenvalue.

(7) Let $A \in M_n(\mathbb{R})$, where n is odd. Then, A has at least one eigenvalue.

(8) The eigenvalues of a symmetric matrix (with real entries) are always real.

(9) An eigenvector may be associated with two distinct eigenvalues.

(10) Let A be a square matrix with real entries only. If λ is a complex eigenvalue of A, so is $\overline{\lambda}$.

Answers.

(1) The given argument contains a logical flaw. First, we do have $p(I) = 0_{M_n}$. The error lies in conflating $p(x) = x - 1$, a general polynomial, with the characteristic polynomial of I. The characteristic polynomial of I is not $x - 1$; instead, it is:

$$p_I(x) = \det(xI - I) = (x - 1)^n.$$

Thus, the polynomial $x - 1$ is *not* the characteristic polynomial of I, and the argument does not apply to Theorem 6.1.2.

(2) **False!** While all nonzero elements of $\ker(f - \lambda I)$ are, in effect, eigenvectors associated with λ, the zero vector is also an element of $\ker(f - \lambda I)$, and it is not considered an eigenvector by definition!

(3) **False!** It is not invertible. Let us show this. Let $\lambda = 0$ be an eigenvalue of a square matrix A, then there exists a non-null vector X such that $AX = 0$. This means that the linear map

associated is not injective, so it is not bijective. Hence, A is not invertible.

(4) It depends on the context. If the matrix is defined on $\mathbb{C}$, then the characteristic polynomial, of degree n, always admits n roots (not necessarily all distinct). This is thanks to the Fundamental Theorem of Algebra.

   If we work with a matrix on $\mathbb{R}$, then the characteristic polynomial of degree n doesn't always admit n roots (distinct or not).

(5) **True.** For a proof, see Exercise 6.3.8.

(6) **False!** Consider the real matrix $A = \begin{pmatrix} 0 & -1 \\ 1 & 0 \end{pmatrix}$. Its characteristic polynomial is given by $\lambda^2 + 1$, which doesn't have roots in $\mathbb{R}$, so A is without real eigenvalues.

(7) **True.** Since A is a square matrix with real entries of odd order, its characteristic polynomial is a polynomial of odd degree. However, and due to the Intermediate Value Theorem, a real polynomial of odd degree always has at least one real root, and eigenvalues correspond to the roots of the characteristic polynomial. Thus, A must have at least one real eigenvalue.

(8) **True.** We accept it without proof here. For instance, and without calculations, the matrix

$$A = \begin{pmatrix} 25796 & 10 & -56 & -223 \\ 10 & 85997 & 12 & 48 \\ -56 & 12 & -58522 & 16 \\ -223 & 48 & 16 & -100023 \end{pmatrix}$$

has only real eigenvalues!

(9) **False!** Let λ and μ be two *distinct* eigenvalues of a linear transformation $f : E \to E$ and suppose there is an eigenvector x (remember that $x \neq 0_E$) such that $f(x) = \lambda x$ and $f(x) = \mu x$. Hence $\lambda x = \mu x$, or $(\lambda - \mu)x = 0$. Since $x \neq 0$, it ensues that $\lambda = \mu$, which contradicts the assumption of λ and μ being distinct.

(10) **True.** Since A has only real entries, the characteristic polynomial of A will have real coefficients only. This is because the characteristic polynomial is given by $\det(A - \lambda I)$, and since A has real entries, the determinant results in a polynomial with real coefficients.

   From basic results in Algebra, it is known that if a complex number λ is a root of a polynomial with *real* coefficients, then its complex conjugate $\overline{\lambda}$ must also be a root of that polynomial.

Therefore, if λ is a complex eigenvalue of A, $\overline{\lambda}$ must also be an eigenvalue of A.

However, things are different when at least one entry is complex. For example, let $A = \begin{pmatrix} 0 & 0 \\ 0 & i \end{pmatrix}$ be defined on $\mathbb{C}^2$. Then, "i" is clearly an eigenvalue of A, while "$\overline{i} = -i$" is not an eigenvalue of A.

6.3. Exercises with Solutions

Exercise 6.3.1. Let A be a diagonal matrix. Determine its eigenvalues.

Solution 6.3.1. The eigenvalues of a diagonal matrix are exactly the elements of its diagonal. Indeed, let A be such a matrix of order n, i.e.,

$$A = \begin{pmatrix} \alpha_1 & 0 & \cdots & 0 \\ 0 & \alpha_2 & \ddots & \vdots \\ \vdots & \ddots & \ddots & 0 \\ 0 & \cdots & 0 & \alpha_n \end{pmatrix}.$$

Then

$$\det(A - \lambda I) = \begin{vmatrix} \alpha_1 - \lambda & 0 & \cdots & 0 \\ 0 & \alpha_2 - \lambda & \ddots & \vdots \\ \vdots & \ddots & \ddots & 0 \\ 0 & \cdots & 0 & \alpha_n - \lambda \end{vmatrix} = 0$$

if and only if λ is any of the values $\alpha_1, \alpha_2, \ldots, \alpha_n$.

For instance, without calculation we know that the eigenvalues of the matrix

$$\begin{pmatrix} -4 & 0 & 0 \\ 0 & 2 & 0 \\ 0 & 0 & 3 \end{pmatrix}$$

are given by -4, 2, and 3.

Remark. We have the same result for a lower triangular matrix, and also for an upper triangular matrix.

Exercise 6.3.2. Let $A = \begin{pmatrix} 4 & 1 \\ 2 & 3 \end{pmatrix}$. Find the eigenvalues and eigenspaces of A.

Solution 6.3.2. The characteristic polynomial is given by:

$$p_A(\lambda) = \det(A - \lambda I) = \det \begin{pmatrix} 4 - \lambda & 1 \\ 2 & 3 - \lambda \end{pmatrix}.$$

Hence,

$$p_A(\lambda) = \lambda^2 - 7\lambda + 10.$$

Solving this quadratic equation yields the eigenvalues of A, which are:

$$\lambda_1 = \frac{7 + 3}{2} = 5, \quad \lambda_2 = \frac{7 - 3}{2} = 2.$$

Let us now find the corresponding eigenspaces. For $\lambda_1 = 5$, we must solve $(A - 5I)u = 0$, i.e.,

$$\begin{pmatrix} 4 - 5 & 1 \\ 2 & 3 - 5 \end{pmatrix} \begin{pmatrix} x \\ y \end{pmatrix} = \begin{pmatrix} 0 \\ 0 \end{pmatrix},$$

which simplifies to:

$$\begin{pmatrix} -1 & 1 \\ 2 & -2 \end{pmatrix} \begin{pmatrix} x \\ y \end{pmatrix} = \begin{pmatrix} 0 \\ 0 \end{pmatrix}.$$

This gives the system of equations:

$$-x + y = 0 \quad \text{and} \quad 2x - 2y = 0.$$

Both equations are equivalent, so $y = x$. Hence, $u = (x, x) = x(1, 1)$, $x \in \mathbb{R}$. An eigenvector is then $(1, 1)$ and the eigenspace corresponding to $\lambda_1 = 5$ is $E_5 = \text{span}(1, 1)$. So, the geometric multiplicity is 1.

For $\lambda_2 = 2$, we need to solve $(A - 2I)v = 0$, i.e.,

$$\begin{pmatrix} 4 - 2 & 1 \\ 2 & 3 - 2 \end{pmatrix} \begin{pmatrix} x \\ y \end{pmatrix} = \begin{pmatrix} 0 \\ 0 \end{pmatrix},$$

which then simplifies to:

$$\begin{pmatrix} 2 & 1 \\ 2 & 1 \end{pmatrix} \begin{pmatrix} x \\ y \end{pmatrix} = \begin{pmatrix} 0 \\ 0 \end{pmatrix}.$$

Thus,

$$2x + y = 0 \quad \text{and} \quad 2x + y = 0.$$

Thus, $y = -2x$. Therefore, the eigenspace corresponding to $\lambda_2 = 2$ is given by $E_2 = \text{span}(1, -2)$, and the geometric multiplicity here is 1 as well.

Exercise 6.3.3. Let $A = \begin{pmatrix} 4 & 1 & 0 \\ 1 & 4 & 1 \\ 0 & 1 & 4 \end{pmatrix}$. Find the eigenvalues and eigenvectors of A.

Solution 6.3.3. We will omit details in this solution. The eigenvalues of A are:

$$\lambda_1 = 4, \quad \lambda_2 = -\sqrt{2} + 4, \quad \lambda_3 = \sqrt{2} + 4$$

as they are roots of $p_A(\lambda) = (4-\lambda)(\lambda^2 - 8\lambda + 14)$. For λ_1, the eigenvector is $v_1 = (-1, 0, 1)$. For λ_2, the eigenvector is $v_2 = (1, -\sqrt{2}, 1)$. Lastly, the eigenvector associated with λ_3 is $v_3 = (1, \sqrt{2}, 1)$.

Exercise 6.3.4. What are the eigenvalues of the matrices

$$A = \begin{pmatrix} 0 & 1 \\ 1 & 1 \end{pmatrix}, B = \begin{pmatrix} 0 & 1 \\ 2 & 2 \end{pmatrix}, \text{ and } C = \begin{pmatrix} 0 & 1 \\ 2 & 0 \end{pmatrix},$$

where $A \in M_2(\mathbb{Z}_2)$ and $B, C \in M_2(\mathbb{Z}_3)$?

Solution 6.3.4. First, recall that $\mathbb{Z}_2 = \{0, 1\}$ and that $\mathbb{Z}_3 = \{0, 1, 2\}$. Also, all calculations are carried out modulo 2 for $\mathbb{Z}_2$, and modulo 3 for $\mathbb{Z}_3$. The characteristic polynomial of A is given by:

$$p_A(\lambda) = \begin{vmatrix} -\lambda & 1 \\ 1 & 1-\lambda \end{vmatrix} = \lambda^2 - \lambda - 1.$$

In $\mathbb{Z}_2$, where $-1 = 1$, this simplifies to

$$p_A(\lambda) = \lambda^2 - \lambda + 1.$$

Since $\mathbb{Z}_2$ contains two elements only, it is straightforward to test each element to see whether it is a root of the characteristic polynomial. Since

$$0^2 - 0 + 1 = 1 \neq 0 \text{ and } 1^2 - 1 + 1 = 1 \neq 0,$$

matrix A does not have any eigenvalues in $\mathbb{Z}_2$.

As for B, readers should easily find that the characteristic polynomial of B is given by $p_B(\lambda) = \lambda^2 - 2\lambda - 2$, or equivalently over $\mathbb{Z}_3$, $p_B(\lambda) = \lambda^2 - 2\lambda + 1$. Now, $p_B(0) = -2 = 1 \neq 0$ and $p_B(2) = 1 \neq 0$, but $p_B(1) = -3 = 0$, which indicates that B has indeed one eigenvalue in $\mathbb{Z}_3$, given by $\lambda = 1$.

Remark. Observe that this 2×2 matrix has only one single eigenvalue (with multiplicity 1), which is rather uncommon. Usually, a 2×2 matrix (with real or complex entries) has either two distinct eigenvalues or one eigenvalue with multiplicity 2 (if repeated), or none. Exercise 6.3.5 presents another surprising or less common fact related to eigenvalues.

In the end, matrix C, whose characteristic polynomial is $\lambda^2 + 1$, has no eigenvalues because this polynomial has no roots in $\mathbb{Z}_3$.

Exercise 6.3.5. Can we find a matrix $A \in M_2$ having four eigenvalues?

Solution 6.3.5. The answer is yes, but the entries must not be in a field! For example, consider $A \in M_2(\mathbb{Z}_{12})$ defined by:

$$A = \begin{pmatrix} 1 & 0 \\ 0 & 3 \end{pmatrix}.$$

All calculations are performed modulo 12. It is plain that

$$p_A(\lambda) = (\lambda - 1)(\lambda - 3).$$

Because $\mathbb{Z}_{12}$ is not particularly an integral domain, we cannot say for certain that $\lambda = 1$ or $\lambda = 3$ are the only outcomes. Indeed, $\lambda = 7$ or $\lambda = 9$ are also roots of the characteristic polynomial of A. Thus, A has four eigenvalues in $\mathbb{Z}_{12}$: 1, 3, 7, and 9.

Exercise 6.3.6. Let A be a square matrix with real entries such that $A^2 + I = 0$, where I is the identity matrix and 0 is the zero matrix. Show that A doesn't admit any real eigenvalues.

Solution 6.3.6. If λ is an eigenvalue of A, then there is a *non-null* vector u such that $Au = \lambda u$. Then,

$$A^2 u = A(Au) = A(\lambda u) = \lambda A u = \lambda^2 u.$$

So,

$$0 = (A^2 + I)u = (\lambda^2 + 1)u \implies \lambda^2 + 1 = 0,$$

which is impossible in $\mathbb{R}$.

Exercise 6.3.7. Prove that the characteristic equation of a real matrix A of order 2×2 can be written as:

$$\lambda^2 - (\mathrm{tr}A)\lambda + \det A = 0.$$

Application: If A is a square matrix of order 2×2 having as eigenvalues the scalars 1 and 2, find its determinant.

Solution 6.3.7. Let

$$A = \begin{pmatrix} a & b \\ c & d \end{pmatrix},$$

where a, b, c, and d are real scalars. Before continuing, we first note that:

$$\mathrm{tr}A = a + d \text{ and } \det A = ad - bc.$$

The characteristic polynomial is given by:

$$(a - \lambda)(d - \lambda) - bc = ad - \lambda d - \lambda a + \lambda^2 - bc = \lambda^2 - \lambda(a+d) + (ad - bc),$$

and this marks the end of proof.

Application: According to the previous question, since 1 and 2 are two real eigenvalues, replacing λ by these two values, we find:

$$\begin{cases} 4 - 2\mathrm{tr}A + \det A = 0, \\ 1 - \mathrm{tr}A + \det A = 0. \end{cases}$$

Solving the system gives: $\det A = 2$ (and as a bonus: $\mathrm{tr}A = 3$).

Remark. More generally, if a matrix A has $\lambda_1, \lambda_2, \ldots, \lambda_n$ as eigenvalues counted with multiplicity, then $\mathrm{tr}A = \lambda_1 + \lambda_2 + \cdots + \lambda_n$. Besides, $\det A = \lambda_1 \lambda_2 \cdots \lambda_n$. However, a full proof requires deeper results that may be beyond our current reach.

Exercise 6.3.8. Let $A, B \in M_n$. Show that AB and BA has the same eigenvalues. Does this result hold if A and B are not square?

Solution 6.3.8. To prove that $\sigma_p(AB) = \sigma_p(BA)$, we treat two different cases: $\lambda = 0$ and $\lambda \neq 0$.

- Let $\lambda = 0$, then
$$\lambda \in \sigma_p(AB) \iff \det(AB) = 0$$
$$\iff \det(A)\det(B) = 0$$
$$\iff \det(BA) = 0$$
$$\iff \lambda \in \sigma_p(BA).$$

- Let $\lambda \neq 0$. If $\lambda \in \sigma_p(AB)$, then there is some $x \neq 0$ such that $ABx = \lambda x \neq 0$. Set $y = Bx$. If $y = 0$, then $A(Bx) = 0$, which is absurd. Thus, $y \neq 0$. Hence,
$$BAy = B(ABx) = B(\lambda x) = \lambda Bx = \lambda y,$$
i.e., λ is an eigenvalue of BA. By interchanging the roles of A and B, the other inclusion can be established, which is left as an exercise for the reader.

If the matrices A and B are not square, still, if the products AB and BA are well-defined, then the set of eigenvalues of AB need not coincide with the set of eigenvalues of BA. As a counterexample, consider

$$A = \begin{pmatrix} 1 & 0 & 0 \\ 0 & 1 & 0 \end{pmatrix} \text{ and } B = \begin{pmatrix} 1 & 0 \\ 0 & 1 \\ 0 & 0 \end{pmatrix}$$

from Exercise 1.3.22. Then,

$$AB = \begin{pmatrix} 1 & 0 \\ 0 & 1 \end{pmatrix} \text{ whereas } BA = \begin{pmatrix} 1 & 0 & 0 \\ 0 & 1 & 0 \\ 0 & 0 & 0 \end{pmatrix}.$$

So, $\sigma_p(AB) = \{1\}$ and $\sigma_p(BA) = \{1, 0\}$. In other words, $\sigma_p(AB) \neq \sigma_p(BA)$.

Remarks.

(1) In the previous counterexample, the fact that the nonzero eigenvalue(s) of AB are exactly the nonzero eigenvalue(s) of BA is not a coincidence. Indeed, if A is an $n \times m$ matrix and B is an $m \times n$ matrix, then the nonzero eigenvalue(s) of AB are exactly the nonzero eigenvalue(s) of BA. Readers can consult Theorem 1.3.22 in [**14**] for a formal proof.

(2) When E is an *infinite-dimensional* vector space and $f, g : E \to E$ are two endomorphisms, then, in general, only the nonzero eigenvalues of $f \circ g$ coincide with the nonzero eigenvalues of $g \circ f$ (the proof can be found in more advanced treatments of operator theory).

Exercise 6.3.9. Show that two similar matrices A and B have the same characteristic polynomial (hence the same eigenvalues).

Solution 6.3.9. By assumption, $T^{-1}AT = B$ for a certain invertible matrix T. Hence, for a scalar λ, we have:

$$\lambda I - B = \lambda I - T^{-1}AT = \lambda T^{-1}T - T^{-1}AT = T^{-1}(\lambda I - A)T.$$

So,

$$\det(\lambda I - B) = \det[T^{-1}(\lambda I - A)T] = \det T^{-1} \det(\lambda I - A) \det T.$$

Since $\det(T^{-1}) = 1/\det T$, we obtain $p_B(\lambda) = p_A(\lambda)$, which completes the proof.

Exercise 6.3.10. Let $A \in M_n$ be a projection, that is, $A^2 = A$. Show that the only possible eigenvalues of A are 0 and 1.

Solution 6.3.10. Let λ be an eigenvalue of A, that is, $Ax = \lambda x$ for some nonzero x. Hence $A^2x = A(Ax) = A(\lambda x) = \lambda^2 x$. But as $A^2 = A$, we see that $A^2x = \lambda x$. In other words, $\lambda^2 x = \lambda x$, which gives $\lambda^2 = \lambda$ since $x \neq 0$. Thus, $\lambda = 0$ or $\lambda = 1$.

Exercise 6.3.11. Find the eigenvalues of the following linear transformations defined from E to E:

(1) $Tf(x) = xf(x)$, $E = C([0, 1], \mathbb{R})$.
(2) $Tf(x) = f'(x)$, $E = C^\infty(\mathbb{R})$.
(3) $Tf(x) = \int_0^x f(t)dt$, $E = C([0, 1], \mathbb{R})$.

Solution 6.3.11.

(1) The transformation T does not possess any eigenvalue. Indeed, let λ be such that $xf(x) = Tf(x) = \lambda f(x)$. If $\lambda \notin [0, 1]$, then $(x - \lambda)f(x) = 0$ gives $f(x) = 0$ since $x \neq \lambda$. Now, let $\lambda \in [0, 1]$. Then $(x - \lambda)f(x) = 0$ implies that $f = 0$ on $[0, 1] - \{\lambda\}$. The continuity of f allows us to obtain that $f = 0$ on the entire $[0, 1]$ as follows: We know that there exists a sequence (x_n) in $[0, 1]$ such that $x_n \to \lambda$, with $x_n \neq \lambda$, so $f(x_n) = 0$ for all n. Thus,

$$f(\lambda) = f\left(\lim_{n\to\infty} x_n\right) = \lim_{n\to\infty} f(x_n) = 0,$$

leading to $f(x) = 0$ for all $x \in [0, 1]$. Accordingly, no real λ can be associated with a nonzero eigenvector. In other words, the set of eigenvalues of T is empty.

(2) In this case, every real number is an eigenvalue for T. Let λ satisfy $f'(x) = \lambda f(x)$ for all $x \in \mathbb{R}$. Solutions of the previous differential equation are given by $f(x) = Ce^{\lambda x}$, where $C \in \mathbb{R}$. Since with each λ, we can associate a nonzero f, the set of eigenvalues for T is the set $\mathbb{R}$.

(3) Let λ satisfy the equation $\int_0^x f(t)dt = \lambda f(x)$. Note that this implies that $f(0) = 0$. Since the left-hand side is continuously differentiable due to the continuity of f, it follows that f is differentiable. Thus, we have $f(x) = \lambda f'(x)$. If $\lambda = 0$, then $f = 0$, which signifies $\lambda = 0$ is not an eigenvalue for T. Now, assume that $\lambda \neq 0$. The previous differential equation has solutions given by $f(x) = Ce^{x/\lambda}$, where $C \in \mathbb{R}$. Since $f(0) = 0$, we obtain $C = 0$, and so $f(x) = 0$ for all x, indicating that T has no nonzero eigenvalues. Therefore, T does not possess any eigenvalues.

Exercise 6.3.12. Let $A, B \in M_n(\mathbb{R})$ be such that $AB - BA = A$.

(1) Show that $A^pB - BA^p = pA^p$ for all $p \in \mathbb{N}$.
(2) Show that if $A^p \neq 0$, then p is an eigenvalue for T, defined in Exercise 3.3.11 (with $A = B$).
(3) Infer that A is nilpotent.

Solution 6.3.12.

(1) We use a proof by induction. The given statement is true when $p = 1$, a hypothesis we have made. Assume now $A^p B - BA^p = A^p$. Left multiplying $A^p B - BA^p = pA^p$ by A, and right multiplying $AB - BA = A$ by A^p imply

$$A^{p+1}B - ABA^p = pA^{p+1} \text{ and } ABA^p - BA^{p+1} = A^{p+1}.$$

Summing up these two equations yields

$$A^{p+1}B - BA^{p+1} = (p+1)A^{p+1},$$

which then settles the desired statement.

(2) If $T(X) = XB - BX$, which is already known to be linear from Exercise 3.3.11, then the result of the previous question may be written as $T(A^p) = pA^p$, which signifies that p is an eigenvalue for T associated with $A^p \neq 0$.

(3) Since the number of distinct eigenvalues is finite due to the finiteness of $\dim M_n(\mathbb{R})$. So, there is a finite number of p such that $A^p \neq 0$. Since p belongs to an infinite set, namely $\mathbb{N}$, then we know that there is at least one $p \in \mathbb{N}$ such that $A^p = 0$; thus, A is nilpotent, as needed.

Exercise 6.3.13. Define a linear transformation $T : M_n(\mathbb{R}) \to M_n(\mathbb{R})$ by $T(A) = A^t$. Find all eigenvalues of T by describing the associated eigenvectors.

Solution 6.3.13. First, remember that here, vectors are, in fact, matrices. We need to find λ such that $A^t = T(A) = \lambda A$ for a nonzero A. Since $A^t = \lambda A$, we obtain $A = (A^t)^t = (\lambda A)^t = \lambda A^t$, whereby $A = \lambda^2 A$, or equivalently, $(1 - \lambda^2)A = 0$. Since $A \neq 0$, it follows by Exercise 2.3.5 that $\lambda^2 = 1$, i.e., $\lambda = \pm 1$, which are the only possible eigenvalues for T.

Let us confirm that these two values are indeed eigenvalues for T by exhibiting explicit eigenvectors. When $\lambda = 1$, any nonzero symmetric matrix, one that verifies $A = A^t$, is an eigenvector; for example, the identity matrix is an eigenvector.

If $\lambda = -1$, then any nonzero skew-symmetric matrix, that is, one that verifies $-A = A^t$, is an eigenvector; for example, consider a non-symmetric matrix B, then set $A = B - B^t$. Then, A is a nonzero skew-symmetric matrix, which can be associated with $\lambda = -1$.

Exercise 6.3.14. Find the eigenvalues of the linear transformation defined by $S : M_n(\mathbb{R}) \to M_n(\mathbb{R})$ defined by $S(A) = A + A^t$.

Solution 6.3.14. To find the eigenvalues for S, consider the equation $A + A^t = S(A) = \lambda A$. Then $A^t = (\lambda - 1)A$, and so $A = (\lambda - 1)^2 A$. So, as above, $(\lambda - 1)^2 = 1$. Thus, $\lambda = 0$ or $\lambda = 2$. As before, choose, as eigenvectors, a nonzero skew-symmetric A in the former case and I in the latter case.

6.4. Exercises without Solutions

Exercise 6.4.1. Let A be a 2×2 matrix with integer entries. Can A have eigenvalues in $\mathbb{Q} \setminus \mathbb{Z}$? Can A have one rational eigenvalue and one irrational eigenvalue?

Exercise 6.4.2. Find two matrices $A, B \in M_n(\mathbb{R})$ such that:

- They have the same characteristic polynomial,
- They share the same eigenvalues (with multiplicities),
- They have the same determinant,
- They have the same trace,

yet they are not similar.

Exercise 6.4.3. Let $\alpha \in \mathbb{R}$. Does the matrix
$$A = \begin{pmatrix} \cos\alpha & -\sin\alpha \\ \sin\alpha & \cos\alpha \end{pmatrix}$$
have eigenvalues?

Exercise 6.4.4. Let $A \in M_n$ be invertible. Show that if λ is an eigenvalue of A, then λ^{-1} is an eigenvalue of A^{-1}.

Exercise 6.4.5. Let λ be an eigenvalue of $A \in M_n(\mathbb{R})$. Can we say that λ^2 is an eigenvalue of $A^t A$?

Exercise 6.4.6. (Cf. Exercise 7.3.14) Let $N \in M_n(\mathbb{R})$ be nilpotent. Find the eigenvalue(s) of N.

Exercise 6.4.7. Set $E = C^\infty(\mathbb{R}, \mathbb{R})$ and $T : E \to E$ be defined by $Tf(x) = f''(x)$. What are the eigenvalues of T?

CHAPTER 7

Diagonalization and Triangularization

7.1. Course Summary

Throughout this chapter, E designates a *finite-dimensional* vector space over a field $\mathbb{F}$ (which is often $\mathbb{R}$ or $\mathbb{C}$) and $f : E \to E$ is an endomorphism.

7.1.1. Diagonalization.

DEFINITION 7.1.1. An endomorphism $f : E \to E$ is said to be diagonalizable if there exists a basis B' in E such that the matrix representation of f with respect to B' is diagonal.

So, matrix $A \in M_n$ is diagonalizable if there is an invertible matrix $P \in M_n$ such that $A' = P^{-1}AP$ is a diagonal matrix.

Below, we have a necessary and sufficient condition for diagonalizability.

THEOREM 7.1.1. *Let $f : E \to E$ be an endomorphism, where $\dim E = n$. Then f is diagonalizable if and only if one of the following equivalent conditions holds:*

(1) $\dim E_1 + \dim E_2 + \cdots + \dim E_p = \dim E$, *where each E_k ($k = 1, 2, \ldots, p$) is an eigenspace associated with an eigenvalue λ_k.*
(2) *E has a basis consisting entirely of eigenvectors of f.*
(3) *The algebraic multiplicity and geometric multiplicity of each eigenvalue of f are equal.*

The following sufficient condition is often useful.

COROLLARY 7.1.2. *If matrix $A \in M_n$ has n distinct eigenvalues, then A is diagonalizable.*

7.1.2. Triangularization.

If a matrix is not diagonalizable, it can still be transformed into an upper (or lower) triangular form via triangularization, with eigenvalues on the diagonal. Though not as direct as diagonalization, this form remains practical for simplifying computations.

DEFINITION 7.1.2. An endomorphism $f : E \to E$ is said to be triangularizable if there exists a basis B' in E such that the matrix representation of f with respect to B' is triangular.

In other words, matrix $A \in M_n$ is triangularizable if there is an invertible matrix $P \in M_n$ such that $A' = P^{-1}AP$ is a triangular matrix.

THEOREM 7.1.3. *Every matrix $A \in M_n(\mathbb{C})$ is triangularizable (in $M_n(\mathbb{C})$). In other words, there exists some invertible matrix P such that:*

$$P^{-1}AP = T,$$

where T is upper triangular with the eigenvalues of A along its diagonal.

Remark. In a more advanced course, the previous theorem can be refined further to obtain a fundamental result in Linear Algebra, namely: Schur's Theorem. It states that for every square matrix $A \in M_n(\mathbb{C})$, there exists a unitary matrix U (i.e., $U^*U = I$), where U^* denotes the conjugate transpose of U as defined in the second remark below Definition 5.1.1, such that:

$$U^*AU = T,$$

where T is an upper triangular matrix with the eigenvalues of A along its diagonal.

7.1.3. Cayley–Hamilton theorem.

The following result, more commonly known as the Cayley–Hamilton theorem, is fundamental in linear algebra, providing a deep connection between matrices and their characteristic polynomials. A proof may be consulted in [8] (Section 10.2).

THEOREM 7.1.4. *Let $A \in M_n(\mathbb{F})$ and $p_A(\lambda) = \det(A - \lambda I)$ be its characteristic equation. Then $p_A(A) = 0_{M_n}$.*

Remark. The Cayley–Hamilton theorem has several important applications. For instance, it allows us to compute powers of square matrices, which is particularly useful when A is not diagonalizable. Another application is that the theorem enables us to find the inverse of an invertible matrix A. This is done as follows: The Cayley–Hamilton theorem states that $p_A(A) = 0_{M_n}$, or equivalently, $a_0 I + a_1 A + \cdots + a_n A^n = 0_{M_n}$, where $a_0 = \det A \neq 0$ (see Theorem 6.1.2). Therefore, we have $-a_0^{-1}(a_1 I + \cdots + a_n A^{n-1})A = I$, which leads to the expression for the inverse:

$$A^{-1} = -a_0^{-1}(a_1 I + \cdots + a_n A^{n-1}).$$

7.2. True or False

Questions. Answer the following questions, providing justifications for your answers:

(1) Each diagonalizable matrix is invertible.

(2) Each invertible matrix is diagonalizable.

(3) Let A and B be two matrices of order n. Then AB is diagonalizable if and only if BA is diagonalizable.

(4) A nilpotent matrix can be diagonalizable.

(5) If A is diagonalizable, then A^2 is diagonalizable.

(6) If A^2 is diagonalizable, then A is diagonalizable.

(7) If A and B are diagonalizable, so is their product AB.

(8) If A and B are diagonalizable, so is their sum $A + B$.

(9) Let $A \in M_n(\mathbb{R})$ be a matrix that obeys $A^2 + A + 2I = 0$. Then A is diagonalizable on $\mathbb{R}^n$.

(10) Let $A, B \in M_n$ be similar and $p(t) = a_0 + a_1 t + \cdots + a_m t^m$. Then $p(A)$ and $p(B)$ are similar.

(11) Let $A \in M_n$. Since $p_A(\lambda) = \det(A - \lambda I)$, it follows that $p_A(A) = \det(A - AI) = 0$. This provides an extremely simple proof of the Cayley–Hamilton theorem, doesn't it?

(12) A diagonalizable matrix is always normal.

Answers.

(1) **False!** Let us exhibit a counterexample. Consider the matrix

$$A = \begin{pmatrix} 1 & 0 \\ 0 & 0 \end{pmatrix},$$

which is already diagonal. It is also plain that A is not invertible.

(2) **False!** Let us give a counterexample. Take the matrix

$$A = \begin{pmatrix} 1 & 1 \\ 0 & 1 \end{pmatrix}.$$

Then A is invertible because $\det A = 1 \neq 0$. The characteristic polynomial of A is given by: $(1 - \lambda)^2 = 0$. Let $u = \begin{pmatrix} x \\ y \end{pmatrix}$ be an eigenvector associated with $\lambda = 1$, then

$$\begin{pmatrix} 1 & 1 \\ 0 & 1 \end{pmatrix} \begin{pmatrix} x \\ y \end{pmatrix} = \begin{pmatrix} x + y \\ y \end{pmatrix} = \begin{pmatrix} x \\ y \end{pmatrix} \implies y = 0, \ x \in \mathbb{R}.$$

Thus, the eigenspace is spanned by $(1, 0)$, so its dimension is 1. Since the algebraic multiplicity of λ is $2 \neq 1$, we deduce that A is not diagonalizable. See Exercise 7.3.1 for another proof.

(3) **False!** Let:
$$A = \begin{pmatrix} 1 & 1 \\ 0 & 0 \end{pmatrix} \text{ and } B = \begin{pmatrix} 0 & 1 \\ 0 & 0 \end{pmatrix}.$$

Then BA is diagonalizable, whereas AB is not.

(4) **False!** See Exercise 7.3.2.

(5) **True.** If A is diagonalizable, we know that $P^{-1}AP = D$ for some invertible matrix P, where D is a diagonal matrix. Hence
$$A^2 = (PDP^{-1})^2 = PDP^{-1}PDP^{-1} = PD^2P^{-1},$$
so $P^{-1}A^2P = D^2$, which means that A^2 is diagonalizable. See Exercise 7.3.10 for a more general result.

(6) **False!** For a counterexample, take $A = \begin{pmatrix} 0 & 1 \\ 0 & 0 \end{pmatrix}$. Then $A^2 = \begin{pmatrix} 0 & 0 \\ 0 & 0 \end{pmatrix}$, which is a diagonal matrix! However, A is not diagonalizable, and a proof may be found in Exercise 7.3.2.

(7) **False!** Consider $A = \begin{pmatrix} 2 & 1 \\ 0 & 1 \end{pmatrix}$ and $B = \begin{pmatrix} 1/2 & 1 \\ 0 & 1 \end{pmatrix}$, which are both diagonalizable for they both have two distinct eigenvalues in a bidimensional space. However, $AB = \begin{pmatrix} 1 & 3 \\ 0 & 1 \end{pmatrix}$ is not diagonalizable. Indeed, since the only eigenvalue of AB is 1, AB being diagonalizable means that $AB = I$, which is untrue. See Exercise 7.3.1.

(8) **False!** Consider $A = \begin{pmatrix} 2 & 1 \\ 0 & 1 \end{pmatrix}$ and $B = \begin{pmatrix} -2 & 0 \\ 0 & -1 \end{pmatrix}$. Then both A and B are diagonalizable because each of them has two distinct eigenvalues in a bidimensional space. Nonetheless, $A + B = \begin{pmatrix} 0 & 1 \\ 0 & 0 \end{pmatrix}$ is nonzero and nilpotent, thus non-diagonalizable (see, e.g., Exercise 7.3.2).

(9) **False!** Assume for the sake of contradiction that there exists an invertible matrix P such that $P^{-1}AP = D$, where D is a diagonal matrix. So, the assumption $A^2 + A + 2I = 0$ reduces to $D^2 + D + 2I = 0$. This means that each α on the diagonal of D necessarily satisfies $\alpha^2 + \alpha + 2 = 0$. Since the latter equation has no solutions in $\mathbb{R}$, the matrix A cannot be diagonalizable.

(10) **True.** Proving it is equivalent to establishing the similarity of $p(A)$ and $p(D)$ in Solution 7.3.10.

(11) **(The given "proof" is) False!** There are several ways to see why the presented argument is incorrect. The main reason is that $p_A(A)$ is a matrix, while $\det(A - AI) = 0$ is a scalar. Therefore, they cannot be equal unless, in the trivial case, $n = 1$.

(12) **False!** As a counterexample, consider $A = \begin{pmatrix} 1 & 1 \\ 0 & 2 \end{pmatrix}$. Then, A is diagonalizable for it has two distinct eigenvalues (on $\mathbb{R}^2$), yet A is not normal.

7.3. Exercises with Solutions

Exercise 7.3.1. Show that a triangular *non-diagonal* matrix A having the same value, e.g., α, on its main diagonal is not diagonalizable. Is this result true for a non-triangular matrix?

Solution 7.3.1. Assume that $T^{-1}AT = D$ for some invertible matrix T, where D is diagonal. Since A is similar to D and A has only α as eigenvalue, it follows that D is necessarily of the form:

$$D = \begin{pmatrix} \alpha & 0 & \cdots & 0 \\ 0 & \alpha & \ddots & \vdots \\ \vdots & \ddots & \ddots & 0 \\ 0 & \cdots & 0 & \alpha \end{pmatrix} = \alpha I.$$

Since the identity matrix I commutes with all square matrices (of the same order), we obtain:

$$A = T(\alpha I)T^{-1} = \alpha I T T^{-1} = \alpha I,$$

which is absurd. Therefore, A is not diagonalizable.

In the end, this result is not true for a non-triangular matrix. Indeed, if, for example, $A = \begin{pmatrix} 1 & 1 \\ 1 & 1 \end{pmatrix}$, which has only "1" on its main diagonal, then $A \neq I$, and A is diagonalizable as readers may check.

Exercise 7.3.2. Show that a nilpotent matrix A is not diagonalizable unless $A = 0$.

Solution 7.3.2. If A is a nilpotent matrix such that $T^{-1}AT = D$ for a certain invertible matrix T, where D is a diagonal matrix, then it may be checked that

$$0 = A^k = (TDT^{-1})^k = \underbrace{(TDT^{-1})(TDT^{-1})\cdots(TDT^{-1})}_{k \text{ times}} = TD^kT^{-1},$$

whereby $D^k = T^{-1}0T = 0$. Thus, $D = 0$, and so $A = TDT^{-1} = T0T^{-1} = 0$.

Exercise 7.3.3. Diagonalize the following matrix:

$$A = \begin{pmatrix} 1 & 2 \\ 2 & 1 \end{pmatrix}.$$

Solution 7.3.3. We have:

$$\det(A - \lambda I) = \begin{vmatrix} 1 - \lambda & 2 \\ 2 & 1 - \lambda \end{vmatrix} = (1 - \lambda)^2 - 4 = \lambda^2 - 2\lambda - 3.$$

Its roots are $\lambda = -1$ or $\lambda = 3$. Let us find their corresponding eigenvectors and eigenspaces. Let u be the eigenvector associated with $\lambda = -1$ and v the eigenvector associated with $\lambda = 3$, respectively. Let $u = \begin{pmatrix} x \\ y \end{pmatrix}$. We have:

$$Au = (-1)u \Longleftrightarrow \begin{pmatrix} x + 2y \\ 2x + y \end{pmatrix} = \begin{pmatrix} -x \\ -y \end{pmatrix} \Longleftrightarrow \begin{cases} 2x + 2y = 0, \\ 2x + 2y = 0, \end{cases}$$

that is, $x = -y$. If E_{-1} is the eigenspace associated with $\lambda = -1$, then

$$E_{-1} = \{\alpha(1, -1) : \ \alpha \in \mathbb{R}\}.$$

In a similar way, if $v = \begin{pmatrix} x \\ y \end{pmatrix}$, then

$$Av = 3v \Longleftrightarrow \begin{pmatrix} x + 2y \\ 2x + y \end{pmatrix} = \begin{pmatrix} 3x \\ 3y \end{pmatrix} \Longleftrightarrow \begin{cases} -2x + 2y = 0, \\ 2x - 2y = 0, \end{cases}$$

i.e., $x = y$.

If E_3 is the eigenspace associated with $\lambda = 3$, then

$$E_3 = \{\beta(1, 1) : \ \beta \in \mathbb{R}\}.$$

Since we have two distinct eigenvalues in a space of dimension 2, we conclude that A is diagonalizable. Additionally, we have:

$$P^{-1}AP = \begin{pmatrix} 3 & 0 \\ 0 & -1 \end{pmatrix} \text{ where } P = \begin{pmatrix} 1 & 1 \\ 1 & -1 \end{pmatrix}.$$

Exercise 7.3.4. Diagonalize the matrix $A = \begin{pmatrix} 0 & 1 \\ -2 & 3 \end{pmatrix}$, then show that

$$A^n = \begin{pmatrix} 2 - 2^n & 2^n - 1 \\ 2 - 2^{n+1} & 2^{n+1} - 1 \end{pmatrix}$$

for all $n \in \mathbb{N}$.

Solution 7.3.4. To diagonalize the matrix A, we need to find a matrix P and a diagonal matrix D such that $P^{-1}AP = D$. The characteristic equation is $\lambda^2 - 3\lambda + 2 = 0$. Thus, $\lambda_1 = 1$, $\lambda_2 = 2$.

Next, we find the eigenvectors corresponding to each eigenvalue by solving the equation $(A - \lambda I)v = 0$, where $v = \begin{pmatrix} x_1 \\ x_2 \end{pmatrix}$.

Let $\lambda_1 = 1$. Solving $(A - \lambda I)v = 0$ yields $\begin{pmatrix} -1 & 1 \\ -2 & 2 \end{pmatrix} \begin{pmatrix} x_1 \\ x_2 \end{pmatrix} = \begin{pmatrix} 0 \\ 0 \end{pmatrix}$.

The eigenvector corresponding to $\lambda_1 = 1$ is $v_1 = \begin{pmatrix} 1 \\ 1 \end{pmatrix}$.

If $\lambda_2 = 2$, we solve $\begin{pmatrix} -2 & 1 \\ -2 & 1 \end{pmatrix} \begin{pmatrix} x_1 \\ x_2 \end{pmatrix} = \begin{pmatrix} 0 \\ 0 \end{pmatrix}$. Thus, the eigenvector corresponding to $\lambda_2 = 2$ is $v_2 = \begin{pmatrix} 1 \\ 2 \end{pmatrix}$.

Consequently, the matrix A is successfully diagonalized as $P^{-1}AP = D$, where $P = \begin{pmatrix} 1 & 1 \\ 1 & 2 \end{pmatrix}$ and $D = \begin{pmatrix} 1 & 0 \\ 0 & 2 \end{pmatrix}$.

In the end, since $A = PDP^{-1}$, it ensues that $A^n = (PDP^{-1})^n$, which simplifies to $A^n = PD^nP^{-1}$. Therefore, for all n

$$A^n = \begin{pmatrix} 2 - 2^n & 2^n - 1 \\ 2 - 2^{n+1} & 2^{n+1} - 1 \end{pmatrix},$$

as wished.

Exercise 7.3.5. Consider the matrix:
$$A = \begin{pmatrix} 0 & -1 \\ 1 & 0 \end{pmatrix}.$$

(1) Show that A is not diagonalizable on $\mathbb{R}$.
(2) Show that A is diagonalizable on $\mathbb{C}$.

Solution 7.3.5.

(1) The characteristic polynomial is $\lambda^2 + 1$. It does not have any real roots. Therefore, A does not have eigenvalues in $\mathbb{R}$. Thus, A is not diagonalizable on $\mathbb{R}$.

(2) The characteristic polynomial is $\lambda^2 + 1$, which, over $\mathbb{C}$, has two distinct roots, namely i and $-i$. The eigenvectors associated with these roots are respectively $\begin{pmatrix} i \\ 1 \end{pmatrix}$ and $\begin{pmatrix} -i \\ 1 \end{pmatrix}$ (they are conjugate because the entries of the matrix are real). So, A is diagonalizable and we have:

$$P^{-1}AP = \begin{pmatrix} -i & 0 \\ 0 & i \end{pmatrix},$$

where $P = \begin{pmatrix} -i & i \\ 1 & 1 \end{pmatrix}$.

Exercise 7.3.6. Prove that the following matrix is diagonalizable:
$$A = \begin{pmatrix} 3 & 0 & 2 \\ -4 & 2 & -5 \\ -4 & 0 & -3 \end{pmatrix}.$$

Solution 7.3.6. We only provide the solution without any accompanying details.

The fact that A has three distinct eigenvalues, namely -1, 2, and 1, makes it diagonalizable. Thus, there is an invertible matrix P such that
$$P^{-1}AP = \begin{pmatrix} 2 & 0 & 0 \\ 0 & 1 & 0 \\ 0 & 0 & -1 \end{pmatrix},$$
where
$$P = \begin{pmatrix} 0 & 1 & 1 \\ 1 & -1 & -2 \\ 0 & -1 & -2 \end{pmatrix}.$$

Exercise 7.3.7. Show that the following matrix is diagonalizable:
$$A = \begin{pmatrix} 4 & -1 & 0 \\ 0 & 3 & 0 \\ 1 & -1 & 3 \end{pmatrix}.$$

Solution 7.3.7. Without excessive detail, the characteristic polynomial is
$$(4 - \lambda)(\lambda - 3)^2,$$
indicating that the matrix has eigenvalues 3 and 4, with 3 having algebraic multiplicity 2.

Let us determine the eigenvectors. Let $u = \begin{pmatrix} x \\ y \\ z \end{pmatrix} \in \mathbb{R}^3$. Then
$$Au = 3u \iff \begin{cases} 4x - y = 3x, \\ 3y = 3y, \\ x - y + 3z = 3z, \end{cases} \iff x = y.$$

The eigenspace is thus given by:
$$E_3 = \{(x, y, z) \in \mathbb{R}^3 : x = y\} = \{(x, x, z), \ x, z \in \mathbb{R}\},$$
So,
$$E_3 = \{x(1, 1, 0) + z(0, 0, 1) : x, y \in \mathbb{R}\}.$$
We realize that the geometric multiplicity of $\lambda = 3$ is 2, which is equal to its algebraic multiplicity.

The eigenspace of the eigenvalue 4 is given by:

$$E_4 = \{(x, y, z) \in \mathbb{R}^3 : y = 0, \ x = z\} = \{x(1, 0, 1), \ x \in \mathbb{R}\},$$

and its dimension is 1. Thus, we have three eigenvectors (counting 3 twice), and since $\dim E_3 + \dim E_4 = 3 = \dim \mathbb{R}^3$, we can conclude that A is diagonalizable. Moreover,

$$P^{-1}AP = \begin{pmatrix} 4 & 0 & 0 \\ 0 & 3 & 0 \\ 0 & 0 & 3 \end{pmatrix},$$

where

$$P = \begin{pmatrix} 1 & 1 & 0 \\ 0 & 1 & 0 \\ 1 & 0 & 1 \end{pmatrix}.$$

Exercise 7.3.8. Is the matrix

$$A = \begin{pmatrix} 1 & 0 & 1 \\ -1 & 2 & 1 \\ 1 & -1 & 1 \end{pmatrix}$$

diagonalizable?

Solution 7.3.8. The characteristic equation is:

$$(1 - \lambda)^2(2 - \lambda) = 0.$$

The eigenvalues of A are 2 and 1, with 1 having algebraic multiplicity 2. For A to be diagonalizable, it is necessary that the dimension of the eigenspace associated with the eigenvalue 1 (its geometric multiplicity) is equal to its algebraic multiplicity, i.e., it must be 2. So, let $u = \begin{pmatrix} x \\ y \\ z \end{pmatrix} \in \mathbb{R}^3$ be such that $Au = 1 \cdot u = u$. Then,

$$Au = u \Longleftrightarrow \begin{cases} x + z = x, \\ -x + 2y + z = y, \\ x - y + z = z, \end{cases} \Longleftrightarrow \begin{cases} z = 0, \\ -x + y + z = 0, \\ x = y, \end{cases}$$

which leads to $x = y$ and $z = 0$, and then the eigenspace is given by:

$$E_1 = \{(x, y, z) \in \mathbb{R}^3 : x = y, \ z = 0\} = \{x(1, 1, 0), \ x \in \mathbb{R}\}.$$

Clearly, $\dim E_1 = 1$, which is not equal to the algebraic multiplicity of the eigenvalue 1. Therefore, A is not diagonalizable.

Exercise 7.3.9. Show that

$$A = \begin{pmatrix} 4 & -1 & 0 \\ 0 & 3 & 0 \\ 1 & -1 & 3 \end{pmatrix}$$

is diagonalizable.

Solution 7.3.9. The characteristic equation is

$$\det(A - \lambda I) = (4 - \lambda)(3 - \lambda)^2 = 0.$$

Thus, the eigenvalues are $\lambda = 4$ with algebraic multiplicity 1, and $\lambda = 3$ with algebraic multiplicity 2.

Now, the eigenspace corresponding to $\lambda = 4$ is $E_4 = \text{span}(1, 0, 1)$, that is, the geometric multiplicity for $\lambda = 4$ is 1.

For $\lambda = 3$, the geometric multiplicity is 2, because it can easily be shown that $E_3 = \text{span}\{(1, 1, 0), (0, 0, 1)\}$. Thus, A is diagonalizable as the algebraic multiplicity of each eigenvalue matches the geometric multiplicity of the same eigenvalue. Accordingly,

$$A = P \begin{pmatrix} 4 & 0 & 0 \\ 0 & 3 & 0 \\ 0 & 0 & 3 \end{pmatrix} P^{-1}, \text{ where } P = \begin{pmatrix} 1 & 1 & 0 \\ 0 & 1 & 0 \\ 1 & 0 & 1 \end{pmatrix}.$$

Exercise 7.3.10. Let $A \in M_n$ be diagonalizable. Let $p(t)$ be a polynomial of degree m with $p(t) = a_0 + a_1 t + \cdots + a_m t^m$. Show that $p(A)$ is diagonalizable.

Solution 7.3.10. Since A is diagonalizable, $T^{-1}AT = D$ for some invertible $T \in M_n$, where D is a diagonal matrix. For the sake of clarity for our readers, we write $T^{-1}AT = D$ as $AT = TD$. Hence,

$$A^2 T = A(AT) = A(TD) = (AT)D = (TD)D = TD^2$$

(this last step appeared elsewhere in this book, and it is worth reiterating here for emphasis). By induction and using the previous approach, readers may also show that $A^m T = TD^m$ for all m. Thus,

$$(a_0 I)T = T(a_0 I), \ (a_1 A)T = T(a_1 D), \ (a_2 A^2)T = T(a_2 D^2), \ldots,$$

$(a_m A^m)T = T(a_m D^m)$, and so

$$(a_0 I + a_1 A + a_2 A^2 + \cdots + a_m A^m)T = T(a_0 I + a_1 D + a_2 D^2 + \cdots + a_m D^m),$$

or equivalently, $p(A)T = Tp(D)$, or even, $T^{-1}p(A)T = p(D)$. Since $p(D)$ is a diagonal matrix by Exercise 1.3.10, $p(A)$ is diagonalizable.

Exercise 7.3.11. Let $A \in M_n$ be a symmetric matrix having only the real α as an eigenvalue.

(1) Show that $A = \alpha I$.
(2) Do we have the same conclusion if A is only normal?
(3) Do we have the same conclusion if the normality of A is dropped?

Solution 7.3.11.

(1) Since α is the only eigenvalue of matrix A, the associated characteristic polynomial, denoted as $p_A(\lambda)$, must be of the form $p_A(\lambda) = (\lambda - \alpha I)^n$, as A is of size $n \times n$. By the Cayley–Hamilton theorem, we have $p_A(A) = (A - \alpha I)^n$. However, $A - \alpha I$ is symmetric due to the realness of α, as stated in Exercise 5.3.2. Therefore, according to Exercise 5.3.4, $A - \alpha I = 0$, or simply $A = \alpha I$.

(2) The answer is positive. Since αI commutes with A, Exercise 5.3.14 states that $A - \alpha I$ is normal. By Exercise 5.3.12, we conclude that $A = \alpha I$.

(3) The answer is no longer positive without the normality of A. For instance, take $A = \begin{pmatrix} 2 & 1 \\ 0 & 2 \end{pmatrix}$, which is not normal. Observe that $A \neq 2I = \begin{pmatrix} 2 & 0 \\ 0 & 2 \end{pmatrix}$, even though 2 is the only eigenvalue of A.

Exercise 7.3.12. Let $a, b, c \in \mathbb{R}$ and consider the matrix:

$$A = \begin{pmatrix} a & b \\ b & c \end{pmatrix}.$$

Prove that A is always diagonalizable regardless of the values of a, b, and c.

Solution 7.3.12. The characteristic polynomial of A is given by

$$\det(A - \lambda I) = \begin{vmatrix} a - \lambda & b \\ b & c - \lambda \end{vmatrix} = \lambda^2 - (a + c)\lambda + ac - b^2,$$

whose discriminant is given by $\Delta = (a - c)^2 + 4b^2$, thereby $\Delta \geq 0$.

- If $\Delta > 0$, then the characteristic polynomial has two distinct roots, and so A is diagonalizable.
- If $\Delta = 0$, which corresponds to $b = 0$ and $a = c$, then the matrix A becomes downright diagonal. To recap, A is always diagonalizable.

Remark. The matrix A in this exercise is symmetric. In fact, any symmetric matrix is also diagonalizable and the transition matrix is orthogonal, but the proof is more complicated and slightly outside the scope of the present audience.

Exercise 7.3.13. Define a linear transformation $T : M_n(\mathbb{R}) \to M_n(\mathbb{R})$ by $T(A) = A^t$. Recall that we already know from Exercise 6.3.13 that T has two eigenvalues, namely $\lambda = \pm 1$. Is T diagonalizable?

Solution 7.3.13. Recall that we already know from Exercise 2.3.43 that the eigenspace corresponding to the eigenvalue 1 is of dimension $n(n+1)/2$, and the eigenspace corresponding to the eigenvalue -1 is of dimension $n(n-1)/2$. Since

$$\frac{n(n+1)}{2} + \frac{n(n-1)}{2} = n^2 = \dim M_n(\mathbb{R}),$$

we conclude that T is diagonalizable.

Exercise 7.3.14. Let $A \in M_n$. Show that A is nilpotent if and only if all its eigenvalues are 0. Infer that if $A^m = 0$ for some $m \in \mathbb{N}$ such that $m \geq n$, then $A^n = 0$.

Solution 7.3.14. Let us continue to represent the linear map associated with A by the same letter A. Assume A is nilpotent, i.e., $A^p = 0$ for a certain $p \in \mathbb{N}$. Let λ be an eigenvalue for A, meaning that $Ax = \lambda x$ for some $x \neq 0$. Hence

$$A^2 x = A(Ax) = A(\lambda x) = \lambda Ax = \lambda^2 x.$$

By induction, we may show that $A^p x = \lambda^p x$, which yields $\lambda^p x = 0$, or merely, $\lambda^p = 0$, as $x \neq 0$. Thus, $\lambda = 0$ is the only possible eigenvalue of A.

To show the backward implication, suppose A has all eigenvalues equal to 0. Then the characteristic polynomial of A is of the form $p_A(\lambda) = (\lambda - \lambda_1)(\lambda - \lambda_2) \cdots (\lambda - \lambda_n) = \lambda^n$. By the Cayley–Hamilton theorem, we can say that $0 = p_A(A) = A^n$, indicating that A is nilpotent.

In the end, if $A^m = 0$, then all the eigenvalues of A are null, and so $A^n = 0$ by the last part of the proof.

Remark. Obviously, the last result is only interesting when $m \geq n$.

Exercise 7.3.15. Let x, y, and z be three functions from $\mathbb{R}$ to $\mathbb{R}$ of the same variable t. Consider the following differential system:

$$\begin{cases} x' = 7x - 3y - 4z, \\ y' = -4x + 6y + 4z, \\ z' = 5x - 3y - 2z, \end{cases}$$

where the differentiation is performed with respect to t.

(1) Write the given system in the matrix form $X' = AX$, where
$$X = \begin{pmatrix} x \\ y \\ z \end{pmatrix}.$$

(2) Diagonalize A.

(3) Solve the given system.

Solution 7.3.15.

(1) We are asked to find the coefficient matrix A, which is clearly given by:
$$A = \begin{pmatrix} 7 & -3 & -4 \\ -4 & 6 & 4 \\ 5 & -3 & -2 \end{pmatrix}.$$

If $X' = \begin{pmatrix} x' \\ y' \\ z' \end{pmatrix}$, then $X' = AX$ is equivalent to the given system.

(2) The eigenvalues are 2, 3, and 6, all of which are distinct from each other. Thus, A is diagonalizable. Upon finding the eigenvectors, we can determine that the transition matrix is as follows:
$$P = \begin{pmatrix} -1 & 1 & 1 \\ 1 & 0 & -1 \\ -2 & 1 & 1 \end{pmatrix}.$$

Thus,
$$P^{-1}AP = D := \begin{pmatrix} 2 & 0 & 0 \\ 0 & 3 & 0 \\ 0 & 0 & 6 \end{pmatrix}.$$

(3) We have:
$$X' = AX \Longleftrightarrow X' = PDP^{-1}X.$$

Hence
$$P^{-1}X' = DP^{-1}X \Longleftrightarrow Y' = DY,$$

where we have set $Y = P^{-1}X$. From a differential equations course, we know that the solution of this system is given by:

$$Y = \begin{pmatrix} \alpha e^{2t} \\ \beta e^{3t} \\ \gamma e^{6t} \end{pmatrix}, \quad \alpha, \beta, \gamma \in \mathbb{R}.$$

Now, we go back to the variable X, where $X = PY$. After performing the necessary calculations, we obtain the solution for X as:

$$X = \begin{pmatrix} x \\ y \\ z \end{pmatrix} = \begin{pmatrix} -\alpha e^{2t} + \beta e^{3t} + \gamma e^{6t} \\ \alpha e^{2t} - \gamma e^{6t} \\ -2\alpha e^{2t} + \beta e^{3t} + \gamma e^{6t} \end{pmatrix}, \quad \alpha, \beta, \gamma \in \mathbb{R}.$$

Exercise 7.3.16.

(1) Show that the matrix:

$$A = \begin{pmatrix} 6 & -2 \\ -2 & 9 \end{pmatrix}$$

is diagonalizable.

(2) Let (u_n) and (v_n) be two sequences defined by

$$\begin{cases} u_0 = 1, \; v_0 = 1, \\ u_{n+1} = 6u_n - 2v_n, \\ v_{n+1} = -2u_n + 9v_n. \end{cases}$$

Calculate the general terms of u_n and v_n in terms of n.

Solution 7.3.16.

(1) The characteristic polynomial of matrix A is given by $\lambda^2 - 15\lambda + 50$, and its roots are 5 and 10. This implies that A is diagonalizable. Therefore, there exists an invertible 2×2 matrix P such that

$$P^{-1}AP = \begin{pmatrix} 5 & 0 \\ 0 & 10 \end{pmatrix}.$$

To find P, we determine the eigenvectors associated with 5 and 10, denoted by u and v respectively. Then,

$$Au = 5u \Longrightarrow u = \begin{pmatrix} 2x \\ x \end{pmatrix} = x \begin{pmatrix} 2 \\ 1 \end{pmatrix}, \quad x \in \mathbb{R};$$

and

$$Au = 10u \Longrightarrow u = \begin{pmatrix} y \\ -2y \end{pmatrix} = y \begin{pmatrix} 1 \\ -2 \end{pmatrix}, \quad y \in \mathbb{R}.$$

Thus,

$$P = \begin{pmatrix} 2 & 1 \\ 1 & -2 \end{pmatrix}.$$

(2) The given system can be written in the following matrix form:

$$\begin{pmatrix} u_{n+1} \\ v_{n+1} \end{pmatrix} = \begin{pmatrix} 6 & -2 \\ -2 & 9 \end{pmatrix} \begin{pmatrix} u_n \\ v_n \end{pmatrix} = A \begin{pmatrix} u_n \\ v_n \end{pmatrix}.$$

As in the previous exercise, we set $\begin{pmatrix} u'_n \\ v'_n \end{pmatrix} = P^{-1} \begin{pmatrix} u_n \\ v_n \end{pmatrix}$, and so

$$\begin{pmatrix} u'_{n+1} \\ v'_{n+1} \end{pmatrix} = \begin{pmatrix} 5 & 0 \\ 0 & 10 \end{pmatrix} \begin{pmatrix} u'_n \\ v'_n \end{pmatrix}.$$

We clearly see that $u'_{n+1} = 5u'_n$ and $v'_{n+1} = 10v'_n$. Therefore, (u'_n) and (v'_n) are two geometric sequences with general terms given by: $u'_n = 5^n u'_0$ and $v'_n = 10^n v'_0$ respectively. Moving back to u_n and v_n, we have:

$$\begin{pmatrix} u'_n \\ v'_n \end{pmatrix} = P^{-1} \begin{pmatrix} u_n \\ v_n \end{pmatrix} \Longleftrightarrow P \begin{pmatrix} u'_n \\ v'_n \end{pmatrix} = \begin{pmatrix} u_n \\ v_n \end{pmatrix},$$

i.e.,

$$\begin{pmatrix} 2 & 1 \\ 1 & -2 \end{pmatrix} \begin{pmatrix} u'_n \\ v'_n \end{pmatrix} = \begin{pmatrix} u_n \\ v_n \end{pmatrix}.$$

However, $u_n = 2u'_n + v'_n$ and $v_n = u'_n - 2v'_n$. So,

$$u_n = 2 \times 5^n u'_0 + 10^n v'_0 \text{ and } v_n = 5^n u'_0 - 2 \times (10)^n v'_0.$$

Now, to find u'_0 and v'_0, we need to solve the system:

$$\begin{cases} 2u'_0 + v'_0 = u_0 = 1, \\ u'_0 - 2v'_0 = v_0 = 1, \end{cases}$$

which then gives $u'_0 = \frac{3}{5}$ and $v'_0 = -\frac{1}{5}$. In fine,

$$\begin{cases} u_n = \frac{3}{5} \times 2 \times 5^n + 10^n \times (-\frac{1}{5}) = 6 \times 5^{n-1} - 2 \times 10^{n-1}, \\ v_n = \frac{3}{5} \times 5^n - 2 \times (10)^n \times (-\frac{1}{5}) = 3 \times 5^{n-1} + 4 \times 10^{n-1}. \end{cases}$$

Exercise 7.3.17. Consider the matrix

$$A = \begin{pmatrix} 1 & 2 \\ 3 & 4 \end{pmatrix}.$$

(1) Calculate A^2 and A^3 using the Cayley–Hamilton theorem.
(2) Again, utilizing the Cayley–Hamilton theorem, show that A is invertible and find its inverse.

Solution 7.3.17.

(1) The characteristic equation is given by $\lambda^2 - 5\lambda - 2 = 0$. According to the Cayley–Hamilton theorem, we have $A^2 - 5A - 2I = 0$. But,

$$A^2 = 5A + 2I = \begin{pmatrix} 5 & 10 \\ 15 & 20 \end{pmatrix} + \begin{pmatrix} 2 & 0 \\ 0 & 2 \end{pmatrix} = \begin{pmatrix} 7 & 10 \\ 15 & 22 \end{pmatrix}.$$

To find A^3, we can write

$$A^3 = A^2 A = (5A + 2I)A = 5A^2 + 2A$$

$$= \begin{pmatrix} 35 & 50 \\ 75 & 110 \end{pmatrix} + \begin{pmatrix} 2 & 4 \\ 6 & 8 \end{pmatrix},$$

thereby

$$A^3 = \begin{pmatrix} 37 & 54 \\ 81 & 118 \end{pmatrix}.$$

(2) Since $\det A = -2 \neq 0$, A is invertible. To determine the inverse matrix A^{-1} using Cayley–Hamilton theorem, we proceed as follows:

$$A^2 - 5A - 2I = 0 \iff 2I = A^2 - 5A$$

$$\iff 2A^{-1} = A - 5I$$

$$\iff A^{-1} = \frac{1}{2}(A - 5I)$$

$$\iff A^{-1} = \frac{1}{2}\begin{pmatrix} -4 & 2 \\ 3 & -1 \end{pmatrix}.$$

Exercise 7.3.18. Using Exercise 7.3.14, show that $A = \begin{pmatrix} 0 & 1 \\ 0 & 0 \end{pmatrix}$ does not have any square root.

Solution 7.3.18. Suppose there is a matrix B such that $B^2 = A$. Hence

$$B^4 = A^2 = \begin{pmatrix} 0 & 0 \\ 0 & 0 \end{pmatrix}.$$

Since B is of size 2×2, Exercise 7.3.14 implies that, in fact, $B^2 = 0$. But, this would mean that $A = 0$, which is impossible. In other words, A does not have any square root.

Exercise 7.3.19. Let A be an $n \times n$ matrix.

(1) Assume that A is defined over the complex field and has n pairwise distinct nonzero eigenvalues. Show that A has exactly 2^n square roots.

(2) Show that this result is violated if A does not enjoy either of the assumptions.

Solution 7.3.19.

(1) A way of establishing this result is via proof by induction. Assume that an $n \times n$ matrix with n pairwise distinct nonzero eigenvalues has 2^n square roots. Let A be an $(n+1) \times (n+1)$ matrix with $n+1$ pairwise distinct nonzero eigenvalues. Since A is diagonalizable, there exists an invertible matrix T such that

$$T^{-1}AT = \begin{pmatrix} \lambda_1 & 0 & 0 & \cdots & 0 \\ 0 & \lambda_2 & 0 & \cdots & 0 \\ \vdots & \ddots & \ddots & \ddots & \vdots \\ 0 & \cdots & 0 & \lambda_n & 0 \\ 0 & \cdots & \cdots & 0 & \lambda_{n+1} \end{pmatrix} = D,$$

where $\lambda_1, \lambda_2, \ldots, \lambda_{n+1}$ are pairwise distinct and nonzero. By the induction hypothesis, the matrix with eigenvalues $\lambda_1, \lambda_2, \ldots, \lambda_n$ on the diagonal has 2^n square roots.

It follows that D has $2 \times 2^n = 2^{n+1}$ square roots, since $\lambda_{n+1} \neq 0$. Therefore, A also has 2^{n+1} square roots.

(2) We present three counterexamples to demonstrate the importance of the three major assumptions.

(a) Define on $\mathbb{R}^2$ the 2×2 matrix $A = \begin{pmatrix} -1 & 0 \\ 0 & 1 \end{pmatrix}$. Then, A has two distinct nonzero eigenvalues, namely 1 and -1. However, A does not have any square roots over the real numbers (why?). This illustrates the importance of working over $\mathbb{C}$.

(b) On $\mathbb{C}^2$, define $A = \begin{pmatrix} 1 & 0 \\ 0 & 0 \end{pmatrix}$. The matrix A has two eigenvalues: 0 and 1. It can also be shown that A has exactly two square roots, which are $B = \begin{pmatrix} 1 & 0 \\ 0 & 0 \end{pmatrix}$ and $C = \begin{pmatrix} -1 & 0 \\ 0 & 0 \end{pmatrix}$. Therefore, the assumption that all eigenvalues are nonzero cannot be disregarded.

(c) Over $\mathbb{C}^2$, let I be the identity matrix, which has the eigenvalue 1 with multiplicity 2. On the other hand, we know that I has infinitely many square roots. Thus, the assumption that the eigenvalues are distinct (pairwise) cannot be simply dismissed.

Exercise 7.3.20. Find all square roots of the matrix:

$$A = \begin{pmatrix} 5 & 4 \\ 2 & 3 \end{pmatrix}.$$

Solution 7.3.20. First, we diagonalize matrix A, omitting many details. Then, A has two nonzero distinct eigenvalues: $\lambda_1 = 7$ and $\lambda_2 = 1$. By Exercise 7.3.19, we know that A has $2^2 = 4$ square roots. Moreover, A is diagonalizable and $P^{-1}AP = D$, where

$$D = \begin{pmatrix} 7 & 0 \\ 0 & 1 \end{pmatrix} \quad \text{and} \quad P = \begin{pmatrix} 2 & -1 \\ 1 & 1 \end{pmatrix}.$$

Also, D has exactly four square roots (cf. Exercise 7.3.19), which are given by

$$E_1 = \begin{pmatrix} \sqrt{7} & 0 \\ 0 & 1 \end{pmatrix}, \quad E_2 = \begin{pmatrix} \sqrt{7} & 0 \\ 0 & -1 \end{pmatrix}, \quad E_3 = \begin{pmatrix} -\sqrt{7} & 0 \\ 0 & 1 \end{pmatrix},$$

and

$$E_4 = \begin{pmatrix} -\sqrt{7} & 0 \\ 0 & -1 \end{pmatrix}.$$

Thus, the four square roots of A are given by:

$$B_1 = PE_1P^{-1}, \quad B_2 = PE_2P^{-1}, \quad B_3 = PE_3P^{-1}, \quad \text{and} \quad B_4 = PE_4P^{-1}.$$

For example, readers can verify that

$$B_1 = PE_1P^{-1} = \begin{pmatrix} 2 & -1 \\ 1 & 1 \end{pmatrix} \begin{pmatrix} \sqrt{7} & 0 \\ 0 & 1 \end{pmatrix} \begin{pmatrix} \frac{1}{3} & \frac{1}{3} \\ -\frac{1}{3} & \frac{2}{3} \end{pmatrix} = \begin{pmatrix} \frac{2\sqrt{7}+1}{3} & \frac{2\sqrt{7}-2}{3} \\ \frac{\sqrt{7}-1}{3} & \frac{\sqrt{7}+2}{3} \end{pmatrix},$$

and $B_1^2 = A$.

Exercise 7.3.21. Show that the following matrices are not diagonalizable:

$$A = \begin{pmatrix} -5 & 9 \\ -4 & 7 \end{pmatrix}, \quad B = \begin{pmatrix} 2 & 1 & 1 \\ 1 & 2 & 1 \\ 0 & 0 & 3 \end{pmatrix},$$

then triangularize them.

Solution 7.3.21. We begin with matrix A. Its characteristic equation is given by: $(\lambda - 1)^2 = 0$, meaning that A has one eigenvalue $\lambda = 1$, with algebraic multiplicity 2.

By solving $(A - I)v_1 = 0$, we find that $v_1 = \begin{pmatrix} 3 \\ 2 \end{pmatrix}$, so the eigenspace corresponding to $\lambda = 1$ is one-dimensional. In other words, the geometric multiplicity is 1, which differs from the algebraic multiplicity. Thus, A is not diagonalizable.

To obtain the basis with respect to which matrix A is triangular, it suffices to complete the vector $(3, 2)$ to a basis of $\mathbb{R}^2$. So, we choose the basis $\{(3, 2), (0, 1)\}$. It only remains to find $f(3, 2)$ and $f(0, 1)$, then write them in terms of $(3, 2)$ and $(0, 1)$. Since 1 is an eigenvalue associated with the vector $(3, 2)$, we have $f(3, 2) = (3, 2) = 1(3, 2) + 0(0, 1)$. We also know that $f(0, 1) = (9, 7)$, and if we write $(9, 7) = x(3, 2) + y(0, 1)$, then we find that $x = 3$ and $y = 1$. Thus, matrix A in this new basis is $A' = \begin{pmatrix} 1 & 3 \\ 0 & 1 \end{pmatrix}$. In fact,

$$A = PA'P^{-1},$$

where $P = \begin{pmatrix} 3 & 0 \\ 2 & 1 \end{pmatrix}$.

Next, we deal with matrix B. Its characteristic equation is:

$$\det(B - \lambda I) = (\lambda - 3)^2(1 - \lambda) = 0.$$

Thus, the eigenvalues are $\lambda = 3$ with algebraic multiplicity 2, and $\lambda = 1$ with algebraic multiplicity 1.

The geometric multiplicity for $\lambda = 1$ is 1 for $E_1 = \text{span}(1, -1, 0)$, while the geometric multiplicity for $\lambda = 3$ is also 1 as $E_3 = \text{span}(1, 1, 0)$. Since the geometric multiplicity of $\lambda = 3$ is different from its algebraic multiplicity, B is not diagonalizable.

To triangularize B, we must complete the free set $\{(1, -1, 0), (1, 1, 0)\}$ to form a basis of $\mathbb{R}^3$. As is customary, we add $(0, 0, 1)$, and we leave it to the reader to verify that $\mathcal{B} := \{(1, -1, 0), (1, 1, 0), (0, 0, 1)\}$ is indeed a basis of $\mathbb{R}^3$. If $f : \mathbb{R}^3 \to \mathbb{R}^3$ is the endomorphism associated with B, then

$$f(0, 0, 1) = 0(1, -1, 0) + 1(1, 1, 0) + 3(0, 0, 1),$$

i.e., $f(0, 0, 1) = (1, 1, 3)$ in the new basis $\mathcal{B}$. Besides,

$$f(1, -1, 0) = 1(1, -1, 0) \text{ and } f(1, 1, 0) = 3(1, 1, 0).$$

Thus,

$$B' = \begin{pmatrix} 1 & 0 & 0 \\ 0 & 3 & 1 \\ 0 & 0 & 3 \end{pmatrix} = P^{-1}BP, \text{ where } P = \begin{pmatrix} 1 & 1 & 0 \\ -1 & 1 & 0 \\ 0 & 0 & 1 \end{pmatrix}.$$

Exercise 7.3.22. Consider the matrix:

$$A = \begin{pmatrix} 0 & 1 & 0 \\ -4 & 4 & 0 \\ -2 & 1 & 2 \end{pmatrix}$$

associated with an endomorphism $f : \mathbb{R}^3 \to \mathbb{R}^3$.

(1) Is A diagonalizable?

(2) Triangularize A.

Solution 7.3.22.

(1) Readers can easily find that $p_A(\lambda) = (\lambda - 2)^3$. If A were diagonalizable, we would have $A = PDP^{-1}$ for some invertible matrix P, but as we have only one eigenvalue (counted thrice), $D = 2I$, which would yield $A = 2I$, and this is impossible. Thus, A is not diagonalizable.

(2) Let us first find the eigenspace associated with $\lambda = 2$. Solving the equation $(A - 2I)u = 0$ gives $u = (x, 2x, z)$, which implies that $E_2 = \text{span}\{(1, 2, 0), (0, 0, 1)\}$. Now, consider the *basis* $\{(1, 2, 0), (0, 0, 1), (0, 1, 0)\}$. We have

$$f(0, 1, 0) = (1, 4, 1) = 1(1, 2, 0) + 1(0, 0, 1) + 2(0, 1, 0),$$

and so $f(0, 1, 0) = (1, 1, 2)$ in the *new basis*. Since $f(1, 2, 0) = 2(1, 2, 0)$ and $f(0, 0, 1) = 2(0, 0, 1)$, the matrix A in this *new basis* is:

$$\begin{pmatrix} 2 & 0 & 1 \\ 0 & 2 & 1 \\ 0 & 0 & 2 \end{pmatrix}.$$

7.4. Exercises without Solutions

Exercise 7.4.1. Let A be a square matrix of order $n \geq 2$ such that $(A - 2I)^5 = 0$ and $(A - 2I)^2 \neq 0$. Is A diagonalizable?

Exercise 7.4.2. Consider the following matrix:

$$A = \begin{pmatrix} a & b \\ c & d \end{pmatrix}.$$

Show that A is:

(1) diagonalizable if $(a - d)^2 + 4bc > 0$;

(2) not diagonalizable if $(a - d)^2 + 4bc \leq 0$.

Exercise 7.4.3. Is the matrix A defined in Exercise 7.3.12 diagonalizable if at least one of a, b, and c is complex?

Exercise 7.4.4. (Refer to Exercise 3.4.10 if needed) Let $A \in M_n$ be a projection. Show that A is diagonalizable. Infer that if $B \in M_n$ is such that $B^3 = B^2$, then B^2 is diagonalizable.

Exercise 7.4.5. Determine for which values of real parameters a and b the matrix:

$$\begin{pmatrix} 1 & b & 0 & 0 \\ 0 & a & b & 0 \\ 0 & 0 & 2 & 0 \\ 0 & 0 & 0 & 2 \end{pmatrix}$$

is diagonalizable.

Exercise 7.4.6. Solve the system:

$$\begin{cases} x_0 = 1, \ y_0 = 2, \ z_0 = -1, \\ x_{n+1} = 3x_n - 4y_n + 2z_n, \\ y_{n+1} = x_n - y_n + z_n, \\ z_{n+1} = x_n - 2y_n + 2z_n. \end{cases}$$

Exercise 7.4.7. Using the Cayley–Hamilton theorem, find the inverse of the following matrix:

$$A = \begin{pmatrix} 0 & 1 & 0 \\ 0 & 0 & 1 \\ 1 & -3 & 3 \end{pmatrix}.$$

Again, using Cayley–Hamilton theorem, determine A^3 and A^4.

CHAPTER 8

Positive Semi-definite Matrices on $M_2(\mathbb{R})$

8.1. Course Summary

DEFINITION 8.1.1. Let $a, b, c \in \mathbb{R}$ and consider the *symmetric* matrix:
$$A = \begin{pmatrix} a & b \\ b & c \end{pmatrix}.$$
Say that A is positive semi-definite if $a \geq 0$, $c \geq 0$, and $\det(A) = ac - b^2 \geq 0$. In symbols, we write $A \geq 0$.

Remark. (Important) We have provided the above definition for 2×2 matrices for simplicity due to the lack, at this level, of specific mathematical tools allowing more general definitions of positive semi-definite matrices. Readers will undoubtedly have the opportunity to see more general and equivalent definitions. Nonetheless, in many, even advanced, cases, and as far as counterexamples are concerned, the 2×2 case is often sufficient.

EXAMPLES 8.1.1. Readers may check that the matrices $\begin{pmatrix} 1 & 1 \\ 1 & 1 \end{pmatrix}$, $\begin{pmatrix} 0 & 0 \\ 0 & 1 \end{pmatrix}$, $\begin{pmatrix} 2 & -1 \\ -1 & 1 \end{pmatrix}$ are positive semi-definite, whereas the matrices $\begin{pmatrix} 0 & 0 \\ 0 & -1 \end{pmatrix}$ and $\begin{pmatrix} 2 & 2 \\ 2 & 1 \end{pmatrix}$ are not.

8.2. True or False

Questions. Answer the following questions, providing justifications for your answers:

(1) Let $A = \begin{pmatrix} 1 & 2 \\ 0 & 2 \end{pmatrix}$. Since $1 \geq 0$, $2 \geq 0$, and $\det A = 2 \geq 0$, A is nonnegative semi-definite.

(2) Let $A \in M_2(\mathbb{R})$ be symmetric. Then
$$A \geq 0 \Longleftrightarrow \operatorname{tr}(A) \geq 0.$$

(3) Let $A \in M_2(\mathbb{R})$ be symmetric. Then
$$A \geq 0 \Longleftrightarrow \det(A) \geq 0.$$

Answers.

(1) Although $1 \geq 0$, $2 \geq 0$, and $\det A = 2 \geq 0$, the matrix A cannot be considered positive semi-definite because it is not symmetric, which is a crucial assumption.

(2) If $A \geq 0$, then by the above definition $\mathrm{tr}A \geq 0$. However, $\mathrm{tr}A \geq 0$ does not necessarily imply $A \geq 0$ (witness the symmetric $A = \begin{pmatrix} 2 & 0 \\ 0 & -1 \end{pmatrix}$).

(3) By definition, $A \geq 0$ already contains $\det A \geq 0$. However, $\det A \geq 0$ does not necessarily yield $A \geq 0$. For example, the symmetric $A = \begin{pmatrix} -1 & 0 \\ 0 & -1 \end{pmatrix}$ has a positive determinant, yet it is not positive semi-definite.

8.3. Exercises with Solutions

Exercise 8.3.1. Let $A \in M_2(\mathbb{R})$. Show that if $A \geq 0$, then $A^2 \geq 0$. Is the converse true?

Solution 8.3.1. Since A is symmetric, so is A^2. Since A is necessarily of the form $A = \begin{pmatrix} a & b \\ b & c \end{pmatrix}$, $A^2 = \begin{pmatrix} a^2 + b^2 & * \\ * & b^2 + c^2 \end{pmatrix}$. Since $a^2 + b^2 \geq 0$, $b^2 + c^2 \geq 0$, and $\det(A^2) = [\det(A)]^2 \geq 0$, we see that $A^2 \geq 0$.

The converse is, however, untrue. For a counterexample, take $A = \begin{pmatrix} 1 & 0 \\ 0 & -1 \end{pmatrix}$, which is not positive semi-definite (as $\det(A) < 0$), but $A^2 = \begin{pmatrix} 1 & 0 \\ 0 & 1 \end{pmatrix}$ is clearly positive semi-definite.

Exercise 8.3.2. Let $A \in M_2(\mathbb{R})$ be positive semi-definite. Show that $A = 0$ if and only if $\mathrm{tr}(A) = 0$.

Solution 8.3.2. When $A = 0$, patently $\mathrm{tr}(A) = 0$, so there is nothing to prove. Assume now that $\mathrm{tr}(A) = 0$, i.e., if $A = \begin{pmatrix} a & b \\ b & c \end{pmatrix}$, then $a + c = 0$. Since $A \geq 0$, we also know that $a, c \geq 0$ and $ac - b^2 \geq 0$. Thus, $a = c = 0$ and $b^2 \leq 0$, which leads to $b = 0$, as well. In consequence, we conclude that $A = 0$, as desired.

Exercise 8.3.3. Let $A \in M_2(\mathbb{R})$ be such that $AA^t - A^t A \geq 0$. Show that A is normal, i.e., $AA^t = A^t A$.

Solution 8.3.3. Since $\text{tr}(AA^t - A^tA) = \text{tr}(AA^t) - \text{tr}(A^tA) = 0$ and $AA^t - A^tA \geq 0$, the preceding exercise says that $AA^t - A^tA = 0$, or $AA^t = A^tA$, as required.

Exercise 8.3.4. Let $A \in M_2(\mathbb{R})$ be symmetric and $B \in M_2(\mathbb{R})$.
(1) When is a diagonal matrix positive semi-definite?
(2) Show that B^tB and BB^t are positive semi-definite.
(3) Show that A^2 is positive semi-definite.

Solution 8.3.4.
(1) A diagonal matrix, of size 2×2 as per our book's convention, is positive semi-definite if both diagonal elements are positive numbers.
(2) First, recall that both B^tB and BB^t are symmetric for any matrix B. If $B = \begin{pmatrix} a & b \\ c & d \end{pmatrix}$, then

$$B^tB = \begin{pmatrix} a^2 + c^2 & * \\ * & b^2 + d^2 \end{pmatrix}, BB^t = \begin{pmatrix} a^2 + b^2 & * \\ * & c^2 + d^2 \end{pmatrix},$$

where we do not need to compute all entries as we will use a simple approach to check the second condition of the definition of positive semi-definiteness. Since $a, b, c, d \in \mathbb{R}$, $a^2 + c^2 \geq 0$ and $b^2 + d^2 \geq 0$. It only remains to check that $\det(B^tB) \geq 0$, but a few moments' thought allows us to write

$$\det(B^tB) = \det(B)\det(B^t) = \det(B)\det(B) = [\det(B)]^2 \geq 0$$

as $\det(B) \in \mathbb{R}$. Thus, $B^tB \geq 0$. Similarly, $BB^t \geq 0$ can be obtained.
(3) Since $A^t = A$, it is seen that $A^2 = AA = A^tA \geq 0$ by the previous question.

Exercise 8.3.5. Let $A, B \in M_2(\mathbb{R})$ be positive semi-definite.
(1) Show that $A + B$ is positive semi-definite.
(2) Show that if $A + B = 0$, then $A = B = 0$.

Solution 8.3.5. Consider on $M_2(\mathbb{R})$, the matrices

$$A = \begin{pmatrix} a & b \\ b & c \end{pmatrix} \text{ and } B = \begin{pmatrix} \alpha & \beta \\ \beta & \gamma \end{pmatrix},$$

where we assume that $a \geq 0$, $c \geq 0$, $ac - b^2 \geq 0$; $\alpha \geq 0$, $\gamma \geq 0$, and $\alpha\gamma - \beta^2 \geq 0$ to ensure the positive semi-definiteness of both A and B. To prove either statement, we need to work with

$$A + B = \begin{pmatrix} a + \alpha & b + \beta \\ b + \beta & c + \gamma \end{pmatrix}.$$

(1) Since $A + B$ is symmetric, the positive semi-definiteness of $A+B$ is equivalent to $a+\alpha \geq$, $c+\gamma \geq 0$, and $\det(A+B) \geq 0$. Since the first two conditions are clearly satisfied, we will only prove that $\det(A + B) \geq 0$. We have

$$\det(A + B) = (a + \alpha)(c + \gamma) - (b + \beta)^2$$
$$= ac + a\gamma + \alpha c + \alpha\gamma - b^2 - \beta^2 - 2b\beta.$$

If $c = 0$, then $ac - b^2 \geq 0$ gives $b = 0$; and when $\gamma = 0$, then $\alpha\gamma - \beta^2 \geq 0$ yields $\beta = 0$. In these cases, $A = \begin{pmatrix} a & 0 \\ 0 & 0 \end{pmatrix}$ and $B = \begin{pmatrix} \alpha & 0 \\ 0 & 0 \end{pmatrix}$, and so, $A + B \geq 0$. Let us therefore assume that $c > 0$ and $\gamma > 0$. Given that $ac - b^2 \geq 0$ and $\alpha\gamma - \beta^2 \geq 0$, we can observe that $ac + \alpha\gamma - b^2 - \beta^2 \geq 0$. Also, since $a \geq b^2/c$ and $\alpha \geq \beta^2/\gamma$, it follows that $a\gamma \geq b^2\gamma/c$ and $\alpha c \geq \beta^2 c/\gamma$. Therefore, and in order to show that $\det(A+B) \geq 0$, it is enough to demonstrate that $b^2\gamma/c + \beta^2 c/\gamma \geq 2\beta b$. This inequality can be further simplified to $\beta^2 c^2 + b^2\gamma^2 \geq 2\beta b\gamma c$, or $(\beta c - b\gamma)^2 \geq 0$, which holds true for all β, c, b, γ. Hence, we conclude that $A + B$ is positive semi-definite.

(2) To prove the second statement, we assume that $A + B = 0$. From this, we have $a + \alpha = b + \beta = c + \gamma = 0$. Given that $a \geq 0$, $\alpha \geq 0$, and $a + \alpha = 0$, we can conclude that $a = \alpha = 0$. Furthermore, since $c, \gamma \geq 0$, we also see that $c = \gamma = 0$. Lastly, with $a = 0$ and $\alpha = 0$, we can deduce that $b^2 = 0$ and $\beta^2 = 0$, implying that $b = \beta = 0$. Consequently, we have $A = B = 0$, as needed.

Exercise 8.3.6. (Cf. Exercise 8.4.1) Let $A, B \in M_2(\mathbb{R})$ be positive semi-definite. Is it true that AB is positive semi-definite?

Solution 8.3.6. The answer is negative. For example, the matrices $\begin{pmatrix} 1 & 1 \\ 1 & 1 \end{pmatrix}$, $\begin{pmatrix} 0 & 0 \\ 0 & 1 \end{pmatrix}$ are positive semi-definite, but

$$AB = \begin{pmatrix} 0 & 1 \\ 0 & 1 \end{pmatrix}$$

is not even symmetric; thus, it cannot be positive semi-definite.

Exercise 8.3.7. Let $A, B \in M_2(\mathbb{R})$ be positive semi-definite. Is it true that $AB + BA \geq 0$?

Solution 8.3.7. First, observe that we already know from Exercise 5.3.2 that $AB + BA$ is symmetric because A and B are, in particular, symmetric. But, $AB + BA$ is not always positive semi-definite. Here is a counterexample: Let

$$A = \begin{pmatrix} 1 & 1 \\ 1 & 1 \end{pmatrix} \text{ and } B = \begin{pmatrix} 0 & 0 \\ 0 & 1 \end{pmatrix},$$

which are positive semi-definite matrices. Then,

$$AB = \begin{pmatrix} 0 & 1 \\ 0 & 1 \end{pmatrix} \text{ and } BA = (AB)^t = \begin{pmatrix} 0 & 0 \\ 1 & 1 \end{pmatrix}.$$

Thus,

$$AB + BA = \begin{pmatrix} 0 & 1 \\ 1 & 2 \end{pmatrix}$$

is not positive semi-definite because $\det(AB + BA) = -1 < 0$.

8.4. Exercises without Solutions

Exercise 8.4.1. Let $A, B \in M_2(\mathbb{R})$ be positive semi-definite. Show that AB is positive semi-definite if and only if $AB = BA$.

Exercise 8.4.2. Let $A \in M_2(\mathbb{R})$ be invertible and positive semi-definite. Show that A^{-1} is positive semi-definite.

Exercise 8.4.3. Let $A, B \in M_2(\mathbb{R})$. Assume that $P^t A P = B$, where P is orthogonal. Show that A is positive semi-definite if and only if B is.

Exercise 8.4.4. Let $A \in M_2(\mathbb{R})$ be symmetric. Show that A is positive semi-definite if and only if its eigenvalues are nonnegative.

Bibliography

1. H. Anton. *Elementary Linear Algebra*, 8th edition, John Wiley & Sons, Inc., 2000.
2. E. Azoulay, J. Avignant, G. Auliac. *Problèmes Corrigés de Mathématiques, DEUG MIAS/SM*, Ediscience (Dunod pour la nouvelle édition), Paris, 2002.
3. E. Azoulay, J. Avignant, G. Auliac. *Les Mathématiques en Licence, 1re année. Tome 1: Cours+exos, MIAS.MASS.SM*, Ediscience (Dunod pour la nouvelle édition), Paris, 2003.
4. E. Azoulay, J. Avignant, G. Auliac. *Les Mathématiques en Licence, 1re année. Tome 2: Cours+exos, MIAS.MASS.SM*, Ediscience (Dunod pour la nouvelle édition), Paris, 2003.
5. E. Boman, F. Uhlig. When is $\frac{1}{a+b} = \frac{1}{a} + \frac{1}{b}$ anyway?, *College Math. J.*, **33/4** (2002) 296–300.
6. S. Dehimi, M. H. Mortad. On the closedness of the range of (fractional) powers of certain classes of possibly unbounded operators, *J. Math. Anal. Appl.*, **539/1** part 2 (2024), Paper No. 128492, 31 pp.
7. A. Denmat, F. Héaulme. *Algèbre Linéaire*, Série: TD, Dunod, 1999.
8. S. R. Garcia, R. A. Horn. *A Second Course in Linear Algebra*, Cambridge Mathematical Textbooks, Cambridge University Press, 2017.
9. J. Germoni. *Best of Algèbre 1ère année*, Dunod, 2001.
10. H. Gianella, R. Krust, F. Taieb, N. Tosel. *Problèmes Choisis de Mathématiques Supérieures*, Springer, 2001.
11. R. N. Gupta, A. Khurana, D. Khurana, T. Y. Lam. Rings over which the transpose of every invertible matrix is invertible, *J. Algebra*, **322/5** (2009) 1627–1636.
12. P. R. Halmos. *Finite-dimensional vector spaces*, Reprinting of the 1958 second edition, Undergraduate Texts in Mathematics. Springer-Verlag, New York-Heidelberg, 1974.
13. P. R. Halmos, *Linear Algebra Problem Book*, The Dolciani Mathematical Expositions, Vol. 16. Mathematical Association of America, Washington, DC, 1995.
14. R. A. Horn, C. R. Johnson. *Matrix analysis*, 2nd edition, Cambridge University Press, Cambridge, 2013.
15. L. Lesieur, Y. Meyer, C. Joulain, J. Lefebvre. *Algèbre Linéaire, Géométrie*, Armand Collin, 1977.
16. M. H. Mortad. *Counterexamples in operator theory*, Birkhäuser/Springer, Cham, 2022.
17. M. H. Mortad. *The Fuglede-Putnam Theory*, Lecture Notes in Mathematics, Vol. **2322**. Springer, Cham, 2022.
18. M. H. Mortad. *Basic Abstract Algebra: Exercises and Solutions*, World Scientific Publishing Co., 2022.

19. M. H. Mortad. *Basic Real Analysis: Exercises and Solutions*, World Scientific Publishing Co., (to appear).

20. S. Lipschutz, M. Lipson. *Schaum's Outline of Linear Algebra*, 4th edition, McGraw-Hill, 2004.

21. Wikipedia. https://en.wikipedia.org/wiki/Normal_matrix

Index

www.ingramcontent.com/pod-product-compliance
Lightning Source LLC
Chambersburg PA
CBHW070214190526
45161CB00002B/79